GCSE

Paul Anderson

Series consultant

Bryan Williams

ornes

Published in 2009 by:
Nelson Thornes Ltd
Delta Place
27 Bath Road
CHELTENHAM
GL53 7TH
United Kingdom

09 10 11 12 13 / 10 9 8 7 6 5 4 3 2 1

A catalogue record for this book is available from the British Library

ISBN 978 1 4085 0412 3

Cover photograph: Getty/Mason Morfit

Page make-up and illustrations by Tech-Set Ltd, Gateshead

Printed and bound in Spain by GraphyCems

Photo Acknowledgements

Alamy/Tony Timmington: 5D; **AP Valves:** 2.11B, 5.3A–F; **Dyson:** 1.5A, 2.3A–D, 2.8C, 2.9B, C; **Fotolia:** p10, p11r, bm, 1.1A–C, 1.2A, 1.3B, 1.4A, 1.7B, C, E, 1.8A, B, 1.9A, B, 1.10A–C, 1.11B, C, 1.12A–E, 2.8A, 2.10A, B, p70, p71, 3B, 3.2B, 3.3B, 3.6C, 3.7B, p87rm, 4B, 4.1B–D, 4.2B, 4.4C, 4.5A, 4.7B, 4.8B, C, 4.11C, 5.2A, 5.2C, 5.7A, 5.8A, 5.10B; **Getty Images:** /**Barry Willis** p88tl, p89r, 4.6A; /**Car Culture** 1.6A, /**FPG** 1.7A, /**George Marks** p11tm, /**Lester Lefkowitz** p7, 1.4B, /**Max Oppenheim** 1.2C; **Innova Systems:** 2.8B, 2.9A; **iStockphoto:** p11bl, tl, 1.4C, 1.6C, 1.7D, F, p38, p39, 2.1C, 2.6A, 3.1C, D, 4.2D, 4.3C, 4.5E, 4.7A, 4.13A, 4.14C, D, p122, 5B, C, 5.1A, 5.2D, 5.9B; **Martyn Chillmaid:** p87lm; **Paul Anderson:** 4.9B, 4.15B, 5.5B; **Princess Yachts:** 4A; **Science Photo Library:** /**Andrew Lambert Photography** p87l, /**Chris Priest & Mark Clarke** 3.4B, /**Lowell Georgia** 4.3A, /**Maximilian Stock Ltd** 3.3A, 5.2B, 5.8B, /**Pascal Goetgheluck** 1.2B, 1.3A, 3.4A, /**RIA Novosti** 5.8C, /**Tony Craddock** p87r; **Syntech:** 5.4A–F.

Contents

3 What can we make it from?   70

4 How will we make it?   88

5 Transferring an engineered product to large-scale production

Nelson Thornes and AQA

Nelson Thornes has worked in partnership with AQA to make sure that this book offers you the best possible support for your GCSE course. All the content has been approved by the senior examining team at AQA, so you can be sure that it gives you just what you need when you are preparing for your exams.

How to use this book

This book covers everything you need for your course.

Learning Objectives

At the beginning of each section or topic you'll find a list of Learning Objectives based on the requirements of the specification, so you can make sure you are covering everything you need to know for the exam.

> **Objectives**
> **Objectives**
> **Objectives**
> **Objectives**
> First objective.
> Second objective.

AQA Examiner's Tips

Don't forget to look at the AQA Examiner's Tips throughout the book to help you with your study and prepare for your exam.

> **AQA Examiner's tip**
> Don't forget to look at the AQA Examiner's Tips throughout the book to help you with your study and prepare for your exam.

AQA Examination-style Questions

These offer opportunities to practise doing questions in the style that you can expect in your exam so that you can be fully prepared on the day.

AQA examination questions are reproduced by permission of the Assessment and Qualifications Alliance.

Visit www.nelsonthornes.com/aqagcse for more information.

What is engineering?

Engineering is a very exciting, creative and interesting subject to study. There is much to learn but most of this can be covered through practical designing and making activities. Engineering involves:

- designing products, through drawing, modelling and testing
- selecting the materials to be used
- planning making activities
- making products using a wide range of different equipment, ranging from hand tools, to machine tools, to automated equipment.

Each of these activities is explained in detail later in the book.

Engineering and a future career

Most people associate engineering with fabrication activities like manufacturing cars and aircraft. However, engineering activities are carried out in a huge variety of industries. Engineers are employed by the chemical industry, food manufacturers, clothing companies, electronics companies – almost every company that makes a product employs or uses engineers.

There is a wide variety of jobs available in engineering. These range from hands-on practical roles, such as machine operators and fitters, to technical roles, such as designers and metallurgists, to line and factory management, to the Directors who run the companies.

GCSE Engineering will provide you with a broad introduction to the subject. It will help you develop practical skills and understanding which will support this wide range of career options.

Structure of the course

GCSE Engineering can be studied as either a single GCSE or as a double award. The double award includes all of the content of the single award and extends the core knowledge that this provides.

GCSE Engineering

The grade for the single award will be awarded as a result of completing two units of work.

Unit 1 is assessed by a one hour written examination. This is worth 40% of the total marks. The examination has two sections. One section will be a product study. You will be given a preparation sheet some weeks before the examination. This will state what type of product the examination will focus on and what information you will need to know. The other section, Manufacturing and Materials, will ask you questions on what you have learned during the course.

Unit 2 is a coursework activity. It is worth 60% of the total marks. This will involve designing and communicating an engineered product and making an engineered product. This will be explained in more detail in the section on Controlled Assessment.

GCSE Engineering (Double award)

The grade for the double award will be awarded as a result of completing four units of work.

Units 1 and 2 are the same as for the single award. Unit 1 is worth 20% of the total marks and Unit 2 is worth 30% of the total marks.

Unit 3 is assessed by a one hour written examination. This is worth 20% of the total marks. The exam has two sections. One section will test your understanding of how technology is applied in engineering. The other section will ask you questions about manufacturing systems.

Unit 4 is a coursework activity. It is worth 30% of the total marks. This will involve developing an engineering design and manufacturing an engineered product. This will be explained in more detail in the section on Controlled Assessment.

Double award material in this book is indicated by a tab at the start of a section or by a vertical rule on the left.

Double award

| Engineering 4852 | **Unit 1: External Assessment (48501)**
Written Paper – 1 hour – 75 marks – 40%
Unit 1 consists of two sections, a Product Study section based on information made available in a Preparation Sheet and a section on Manufacturing and Materials | plus | **Unit 1: Internal Assessment (48502)**
Controlled Assessment – Approx 40 hours – 90 marks – 60%
Designing and Communicating/ Manufacturing an Engineering Product
A portfolio/one project for single award, Mechnical, Pneumatic or Electronic |

| Engineering (Double Award) 4854 | **Unit 1: External Assessment (48501)**
Written Paper – 1 hour – 75 marks – 20%
Unit 1 consists of two sections, a Product Study section based on information made available in a Preparation Sheet and a section on Manufacturing and Materials

Unit 2: Internal Assessment (48502)
Controlled Assessment – Approx 40 hours – 90 marks – 30%
Designing and Communicating/ Manufacturing and Engineering Product
A portfolio/one project for single award, Mechanical, Pneumatic or Electronic | plus | **Unit 3: (48503)**
Written Paper – 1 hour – 75 marks – 20%
Unit 3 consists of a section on Application of Technologies and a section on Manufacturing Systems

Unit 4: (48504)
Controlled Assessment – Approx 40 hours – 90 marks – 30%
Developing an Engineering Design/ Manufacturing an Engineering Product
A portfolio/two projects or one project combining two technologies (Mechanical, Pneumatic or Electronic) for Double Award |

A *Engineering coursework*

What is this book about?

This book has been written for students studying GCSE Engineering. Its aim is to support your effort towards achieving an excellent grade in this subject. It covers all of the technical content that you will need to know to be successful and provides background for the development of your practical engineering skills. It includes clear concise explanations, descriptions and examples to show the engineering principles in use in real situations. It also provides examples of examination-style questions to help you develop examination skills and support your revision.

Controlled Assessment

The controlled assessment tasks in this book are designed to help you prepare for the tasks your teacher will give you. The tasks in this book are not designed to test you formally and you cannot use them as your own controlled assessment tasks for AQA. Your teacher will not be able to give you as much help with your tasks for AQA as we have given with the tasks in this book.

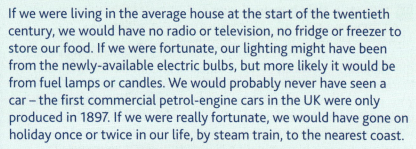

1 Engineering for success

Objectives

Overview how engineered products have affected our life and society.

Explain why new engineered products are created.

If we were living in the average house at the start of the twentieth century, we would have no radio or television, no fridge or freezer to store our food. If we were fortunate, our lighting might have been from the newly-available electric bulbs, but more likely it would be from fuel lamps or candles. We would probably never have seen a car – the first commercial petrol-engine cars in the UK were only produced in 1897. If we were really fortunate, we would have gone on holiday once or twice in our life, by steam train, to the nearest coast.

If we consider how things have changed, we will appreciate how developments in engineering and technology have had a huge impact on our life and our society. We take these developments – the phones, music systems, transport – for granted on a daily basis. They are all the products of engineering. In this chapter, we will investigate how developments in engineering affect the products we use and our society. To do this, we first need a general understanding of why products are created.

■ Why engineered products are created

The starting point for engineering developments is the identification of a need. For example, this could be the need for food storage, to ensure that there is food available every day. In industry, this need is normally something for which users are prepared to pay to have a solution.

The main function of a designer is to create solutions to needs. In this case, they may investigate the different technologies available for food storage and identify that if food is cooled it lasts longer. If this appears to be the best potential solution, they will design a product that provides a cool area for food storage.

To make the solution, engineers may use existing materials or create new or improved materials. Manufacturing processes will be developed and used to change these materials into the product.

After the first products have been manufactured to satisfy the need, the experience gained may allow improvements to the materials and manufacturing processes. This could include the use of computers to improve the design of products or control the manufacturing processes. These developments in turn will allow the design and manufacture of more complicated solutions that better satisfy the need.

Initially, the focus may be on the needs of 'local' users. However, driven by potential rewards, this approach could also be used to satisfy similar needs across the world. Global approaches to producing the solutions to common needs will be developed, maximising the potential rewards for satisfying the needs or making the rate of progress even faster.

In this way, one simple need can lead to a spiral of improvement in engineering technology. However, as the initial need is satisfied, further needs may develop and priorities may change. For example, users may decide they need additional facilities from their cooling units, such as the ability to create ice cubes, or they may need reduced operating costs and power consumption. Needs may also change to reflect the broader impact of the product in the environment. For example, in the case of food storage, until recently the cooling mechanisms in fridges typically contained chemicals that could damage the environment when the fridge reached the end of its usable life. The increasing priority of the need to prevent environmental damage meant that alternative approaches had to be developed for the cooling method. These changing needs may require the frequent redesign of the solution, so that it is able to satisfy all of them.

This chapter will investigate the impact of developments in each of the areas touched upon above in more detail, starting with the impact of developments in materials.

Starter activity

For an electrical product used in the home, create a list of all the different needs that it is used to satisfy. Try to include needs relating to what the product is used for, how it is made and how it is maintained.

AQA Examiner's tip

When looking at the impact of a technology, ensure that you know what is affected (the way a product is made, the environment). Make sure that you have considered all the people who may be affected (users, manufacturer and other people).

A   *Food storage systems*

The impact of developments in materials on engineered products (1)

■ How do materials influence product design?

The term 'materials' covers both the substances that products are made from and the components that they might be made from. Materials are one of the major constraints that have to be considered by the designer and engineer. That means that they limit what can be made. For example, a designer of scaffolding for use on building sites (Photo **A**) might calculate that a piece of scaffolding needs to be able to support the weight of three workers. Using the strength of the material, the designer will calculate what size and shape the material used for the scaffolding frame needs to be. Further, the materials that a product is to be made from can also limit how a product can be made. Some materials can only be made using certain processes. For example, if the scaffolding is made from metal it could not be made by vacuum forming.

The development of materials with improved properties can reduce the impact of the constraint, giving more flexibility in design. These developments can result from either the development of new materials, which will be discussed in this section, or from the improvement of existing materials, which will be discussed in the next section.

■ The development of new materials

The development of new materials with improved properties, or new combinations of properties, can lead to solutions to problems where no solution was previously possible. This can lead to completely new products and markets. The case study on body armour shows one example of this.

However, more commonly the main driving force behind the development of new materials is to improve how well the needs of customers or users are met by an existing application. For example, there has been a continuing need for small lights to show when the switches on control panels are turned on. Traditionally this need was met by using bulbs. In the 1990s new materials were identified that could be used in Light Emitting Diodes. These offered much lower power consumption, with much longer working lives, at lower cost. LEDs are now routinely used as switch indicators instead of bulbs.

A common mistake is to underestimate how many new materials and components are being developed. Hundreds are developed every year. Of particular note currently are 'composite' materials and 'smart' materials. These are discussed in detail in Topic 3.3 and Topic 3.4. 'Composites' have already found many applications, ranging from dental fillings to high-performance aircraft parts. 'Smart' materials have just started to find their first commercial uses. The number of applications for both types of material is expected to grow significantly over the next decade.

Objectives

Explain how the development of new materials can lead to new products and improvements in the performance of existing engineered products.

Show how developments in materials have affected an engineered product.

A *Scaffolding*

Materials for body armour

In medieval times, warriors could be protected from injury from swords and arrows by wearing metal armour, Photo **B**. This typically comprised thick, heavy sheets of metal, often on top of material made from linked metal rings. Unfortunately, as firearms were developed, this armour proved to be ineffective. To stop a bullet, it had to be so heavy that the wearer was unable to walk!

Most practical solutions to body armour after this time involved lots of padding. These did not stop bullet penetration, but helped to reduce injuries from shrapnel and debris.

The plastic Kevlar was developed in 1965 and first used in the 1970s. Compared with the materials that were available before, it was much more difficult to break. Weight-for-weight, it is five times stronger than steel. It led to the development of 'bullet proof' body armour that could not be penetrated by bullets. This is now widely used by the military and police, Photo **C**. Considering just one of its uses – by police in the US – it has been documented as saving over 2000 lives.

B *Medieval armour*

C *Modern body armour*

Information

Some of the new materials developed during the last century

- Nylon, synthetic rubber, acrylic, polyethylene
- Glass reinforced plastic (GRP)
- Transistors
- Microprocessors
- Plasma and LCD displays
- Shape memory alloys
- Biomaterials and dental composites
- Quantum tunnelling composites

Key terms

Materials: the substances or components that products are made from.

Constraints: things that limit what you can make.

Process: an operation that changes the size, shape or condition of a material.

Activity

Select one of the new materials or components developed during the last century from the examples in the Information box. Using the internet, find examples of products that contain this material or component. Identify what material or component was replaced by the new material and the advantages that were obtained by using the new material.

Summary

Materials are a major constraint that must be considered by the designer and engineer.

The development of materials with improved properties, or new combinations of properties, can have a significant impact on the design of products.

AQA Examiner's tip

You should be able to list a range of new materials and some of the applications for which they may be suitable.

1.2 The impact of developments in materials on engineered products (2)

■ The effect of improving the properties of existing materials

It is easy to understand how the development of a completely new material can lead to a new product or a big change in the **performance** of an existing product. However, over the lifetime of a product, its performance may also be improved due to small improvements in the materials being used. Whilst the individual changes in performance from these improvements are normally smaller than using a completely new material, they typically occur much more frequently. For example, when washing powder was first developed, it replaced soap bars. This 'step-change' was achieved through the use of new materials; in this case the ingredients of the washing powder. However, year after year, trips to the supermarket will reveal a huge number of 'new' or 'improved' washing powders, each with small changes to their ingredients. The products currently available, including all the years of small improvements, perform much better than the original development.

This ongoing process of improving the properties of existing materials can affect the design of products in two main ways. They can extend how well a product meets its needs or address additional needs, as shown in the case study on spectacle frames. Alternatively, they can allow changes in the **form** of the product. This may arise due to changes in the form of the material, as shown in the case study on mobile phones, or improvements in the properties of the materials

Objectives

Explain how the development of existing materials can lead to improvements in the performance of existing engineered products.

Show how developments in materials have affected an engineered product.

Case study

Spectacle frames

Spectacle frames have always been susceptible to accidental damage, for example when someone drops them or sits on them. Historically, designers have tried to address this by a variety of different approaches, such as using stronger materials, or 'thicker' frames, Photo **A**, or by including spring-loaded hinges for the arms. Although they reduce the risk of damage, these solutions can affect the appearance of the frames. One alternative approach was to use a type of metal alloy developed in the 1960s, which had the ability to return to a product's original shape after it had been bent.

A *Plastic spectacle frames*

However, the flexibility of the material was limited, which meant that it could still be prone to some damage. Between 1988 and 1993 a number of improved alloys were developed, with much greater flexibility. The use of these alloys for spectacle frames has now grown significantly, Photo **B**. They allow simple designs which can be easier to make and address the need of being more resistant to accidental damage.

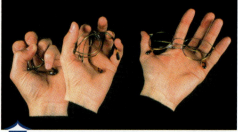
B *Spectacle frames made from shape memory alloy*

used. For example, aircraft frames are normally made from metal **alloys**. Small changes to the mixtures of metals used to make these alloys have led to increased strength. This has allowed designers to use less material in the frame. This has reduced the weight of the aircraft, increasing its efficiency. It also means that less materials are needed, which is more environmentally friendly.

The development of mobile phones

The first commercial mobile phone service was launched in Japan in 1978. The first mobile phones were large and bulky, Photo **C**. The handset was often bigger and heavier than a house brick. The size and weight was necessary due to the electrical components used to make it and also the battery. The only function they provided was to make and receive calls.

Over time, improvements in electronics and battery technology have shrunk both types of component in size. For example, an electrical product that in 1970 was the size of a wardrobe can now be smaller than a matchbox. In addition, the smaller components mean products of a similar size can have additional capabilities. Mobile phones can now send texts, take pictures and play music, addressing additional user needs.

The limiting factor on the size of a mobile phone is now no longer the components – it is the need to be big enough so that it can be operated by hand.

Modern mobile phone

Early mobile phone

 C *How current mobile phones compare with early models (same scale)*

Key terms

Performance: how well a product meets the needs of its users.

Form: the size, shape and condition of a piece of material, product or part.

Alloy: a mixture of two or more metals.

Activity

Using the internet, find examples of how the design of television sets has changed over time. You should identify the causes of each change, including both the electrical components and the materials used for the box or casing.

Summary

Improvements to the original materials may occur several times over the life of a product. This may allow products to meet existing needs better or to meet additional needs.

Changes in the form of the materials being used and improvements in the properties of the materials can both allow changes in the design of a product.

AQA Examiner's tip

You should be able to explain, using examples, how developments in components have affected the design of electronic products.

1.3 The impact of process developments on products (1)

How does process development affect the design of products?

Processes are manufacturing operations that change the size, shape or condition of a material or part in some way. Machining, assembly and heat treatment activities are all processes. The designer has to consider which processes might be used to make products to his design. He may have to modify the design to take account of the process **constraints**. These constraints are typically due to either process capability or the cost of using a process.

Capability means the ability to make the product to the level of accuracy that the designer has calculated is necessary. Some processes are only capable for use with certain **forms** of material. For example, vacuum forming can be used for sheets of thermoplastic, but it does not have the capability to shape steel plate. Even if a process can manufacture parts using the material, it may still not be capable of achieving the required accuracy.

It is almost certain that there are other processes that could be used to make the product. However, these may cost significantly more to buy or to use. Cost constraints may prevent the buying of new equipment to make a product, limiting the process choices to those that are already available in a factory. Cost constraints may also limit the amount of labour and machine time to manufacture a product.

The development of manufacturing processes can improve the capability of processes or reduce the cost of manufacturing operations. This can reduce the impact of process constraints, giving more flexibility in design and manufacturing. In broad terms, there are two approaches to process development: creating new methods of manufacturing, which will be covered in this topic, or improving existing technologies, which will be covered in the following topic.

Creating new methods of manufacturing

Unlike the development of new materials, the creation of completely new methods of manufacturing is rare. Those which are developed normally arise from academic research at universities or company research centres.

One of the most important reasons for the development of new processes is to manufacture new materials, because the existing processes are not capable. For example, the development of thermoplastics led to practical challenges, such as how products made from different types of plastic could be permanently joined together. One of the solutions developed in the 1960s was 'ultrasonic welding'. In this process, ultrasonic vibrations are passed down a plastic part. This causes friction where it contacts other parts, which creates heat.

Objectives

Explain how processes can be a constraint during the design of a product.

Describe some of the reasons why new manufacturing methods may be developed.

This causes the contacting surfaces to melt and they fuse together. This process is now commonly used in products with small parts, such as mobile phones and disposable medical instruments.

Similarly, abrasive water jet cutting was developed to cut materials which could be damaged by heat, Photo **A**. Many industrial processes use heat to melt the materials to be cut and therefore are not suitable. This method uses water mixed with an abrasive material, such as sand. It uses very high pressure to blast this mixture very accurately at the material to be cut. This wears away the material on the cut line. The major advantage of this process is that the material being cut remains cold; it can even be used for flammable materials. It can be used for composites that contain plastics. A disadvantage is that the water pressure is so high that the tool must be held and guided by an automated machine.

Some manufacturing methods are developed as a result of technology advances in other areas. For example, the development of lasers led to applications being identified where they could be used to cut or weld metal and plastic parts, Photo **B**. An advantage of lasers is that they are very accurate and can cut very quickly. However, the equipment is very expensive, which means that it is only normally cost effective if large quantities of parts are being made.

A   *Water jet cutting of cake*

B   *Laser cutting of steel sheet*

Activity

Select one of the new methods of manufacturing outlined above. Using the internet, find examples of materials and products that are manufactured using this approach.

Key terms

Process: an operation that changes the size, shape or condition of a material.

Constraints: things that limit what you can make.

Form: the size, shape and condition of a piece of material, product or part.

Summary

Designers have to consider the process constraints to ensure that their design can be made. Process development can help to reduce these constraints by increasing process capability or reducing the cost of using them.

The development of completely new manufacturing methods is rare. However, these may be necessary to provide the capability needed to manufacture new materials. They may also result from importing new technologies from other areas.

AQA Examiner's tip

You should be able to explain why different methods of manufacturing may be needed.

1.4 The impact of process developments on products (2)

Process improvement

Most manufacturing processes are carried out using machines. If these are controlled by an operator, they are called 'manual machines'. Up to the 1950s, all machining processes were carried out on manual machines. They are still produced and are in common use today, as they are suitable for many applications. The accuracy of these processes depends on the skill of the machine operator.

It is possible to make some improvements to manual machining processes by the use of **jigs** and **fixtures**. These can reduce the time required to load parts into machines and help to improve the accuracy of machining. The cost of manufacturing these simple devices means that they are normally only made when making more than one product. They are explained in detail in Topic 4.12.

In the 1940s, methods of using electronics to control machining processes were developed, called **Numerical Control (NC)**. The machining process being carried out on a manual machine and its equivalent NC machine is the same; the only difference is the way that the machining process is controlled. The first NC machine was used in industry in 1952. NC machines could produce difficult shapes over and over again, with greater accuracy than skilled human operators. They could also carry out machining faster than manual machines. In 1957, some of the earliest industrial computers were used to replace some of the electronics, giving rise to **Computer Numerical Control (CNC)**. As the performance of computers has improved over time, the accuracy and speed of these machines has further improved. The main disadvantages of using CNC machines are that they cost more than manual versions and need to be programmed. This means that a large quantity of parts has to be made before the cost of the product being made is reduced.

Practically every manufacturing company now uses some CNC machines. For some companies manufacturing large quantities of parts, every machine may be CNC. They have led to greater accuracy of machining, the ability to make more parts and reduced product costs for parts made in large quantities.

Objectives

Explain how the use of electronic and computer controls can improve the performance of a process.

Describe how process developments have affected an engineered product.

Key terms

Jig: a work holding or positioning device that is not fixed to the machine bed.

Fixture: a work holding or positioning device that is fixed to the machine bed.

Numerical Control (NC): using numerical data to control a machine by electronic means.

Computer Numerical Control (CNC): using numerical data to control a machine

Case study

Manufacture of a cylinder block

The cylinder block is the largest component in a petrol or diesel engine. It houses the cylinders and forms the bottom part of the combustion chambers; the top part is formed by the cylinder head. The combustion chambers are where the fuel explodes, pushing the pistons to create movement. If there are any gaps the pressure from the exploding fuel will escape from the engine and not drive the pistons. This means that the surfaces where the two parts meet need to be flat so they fit together very well, Photo **A**.

A *Machined surface on a cylinder block*

Although there have been developments in design and the materials used, the types of process used to manufacture the cylinder block have been the same for over 100 years. For the flat surface where the cylinder block and head meet, called the deck, one of the key processes used is milling.

Originally, this was carried out using manual machines, similar to the type shown in Photo **B**. Jigs and fixtures were used to position the cylinder block accurately. For a four cylinder block, the time to mill the deck was typically between 15 and 60 minutes. The machine had to be controlled by a skilled operator, constantly paying attention to the machining.

The development of CNC milling machines, of the type shown in Photo **C**, has reduced the time to mill the deck. Some recent CNC machines have reduced the time to less than one minute. The operator can run two or more CNC machines at the same time. Together with the faster machining time, this significantly reduces the labour cost per part. Further, the CNC machines are more accurate – the variations in the 'flatness' achieved using the CNC machines can be less than 50% of those typically achieved using a manual machine.

B   *Manual milling machine*

C   *Computer-controlled milling machine*

Activity

Using the internet, find examples of how the design of kettles has changed over time. Investigate the different processes that have been used to make kettles and explain how these influence the design.

Summary

One method of improving the capability and reducing manufacturing costs with manual machines is to use jigs and fixtures.

NC machines use electronics to control machines using numerical data. CNC machines are NC machines that are controlled by computers.

CNC machines cost more than manual machines. However, they can typically machine parts much faster and more accurately.

AQA Examiner's tip

You should be able to explain the advantages of using computer-controlled machines and robots.

The impact of developments in information technology on the product development process

Information technology and the role of the designer

Information technology has revolutionised product development. This can be demonstrated by considering its effect on the role of the designer.

Before information technology was available, designers would produce drawings by hand. A separate team would build prototypes and models to test the ideas. The designer would review the test results and modify the design to improve it, producing more new drawings by hand. Another prototype would be made. For a complex product, it could take several months and thousands of labour hours going through this process before a final product design was ready to be passed to manufacturing.

Using information technology, the designer produces the drawings using **Computer Aided Design (CAD)** software. The designer can test the ideas immediately using computer models, as shown in Photo **A**. The drawings can be modified quickly, by editing the originals using the CAD software. The labour time and materials cost to make lots of different prototypes is greatly reduced, significantly reducing the total cost of product development. The whole process can be carried out in days, rather than months.

The use of information technology can also allow more creative designs. The designer can use the models to test out radical design ideas, without the expense of having to make prototypes.

The reasons for using information technology

The impact of information technology is not limited to product design. It can be used at every stage of the product development process, Table **B**. Many benefits have been achieved by using information technology in every stage of product development.

As noted above, the cost to develop new products is reduced. This cost is normally divided between the products being made, which means that the product cost is reduced. Where products are being made in large quantities, the cost per product can often be further reduced by using computer-controlled machines. These can carry out machining processes faster than manual machines, as explained in Topic 1.4. They can also be more accurate, reducing costs for scrap and rework.

Products move from being an initial idea to a finished item, ready for sale, much faster. In addition to the reduction in design time noted above, the availability of design information through a computer network can allow the manufacturing team to obtain tools and materials and to plan some parts of the manufacturing process while the fine details of the design are still being finalised. This is called **concurrent engineering**. It means that the changing needs of customers can be met quickly, which can give an advantage over

A Modelling of airflow within the component

competitors. This also makes it easier to make occasional 'small' changes to products to make them more attractive to customers. For example, this could be a 'limited edition' special model of a car or a child's lunchbox printed with characters from a newly released film.

B *Examples of how information technology can be used during product development*

Activity	How it was carried out ...	Could now be carried out ...
Researching customer needs and the performance of competitors products	Surveys, libraries, reference books, sales brochures, product analysis	E-mail surveys, using information sources on the internet
Creating drawings	Hand-drawn by draughtsmen	Drawn using computer software
Testing ideas	Making prototypes and physical models	Using virtual models (computer software)
Sending drawings to manufacturing	Copying drawings and sending them in internal (or external) post	E-mail
Sharing information with other people in the company	Internal memos	Internal e-mail
Finding suppliers for materials	By visiting companies and trade shows	Using internet search engines
Placing orders for materials	Telephone and post	E-mail
Paying suppliers for materials	Sending cheques by post	Electronic bank transfers using the internet
Planning and scheduling manufacturing	Planning department prepared plans by hand	Using specialist computer software
Moving materials around the factory	Forklift trucks	Robot vehicles
Loading parts into machines	By machine operators	By robots and automated machines
Making parts	Using manual machines	Using computer-controlled machines
Assembling the product	By human workers	By robots
Measuring the product	By hand by quality engineers	Using computer-controlled sensors
Packing the product in boxes	By hand by despatch staff	By robots
Tracking the product	Noting serial numbers	Scanning barcodes
Keeping product records	Filing cabinets full of information	Database on computer
Advertising products	Adverts in mass media	Websites
Keeping records of customers	Filing cabinets full of information	Database on computer

Activity

Produce a working drawing of a table by hand using drawing equipment. Produce the same design using a CAD software package. Compare the results from the two methods.

Summary

Information technology can be used at every stage of the product development process. It can also support concurrent engineering within the process.

Before information technology, every drawing was created by hand and prototypes had to be made to test every design idea. Designers can now produce drawings and models using CAD software.

The use of information technology has led to reduced product costs and allowed products to move from being an initial idea to a finished item much faster.

Key terms

Computer Aided Design (CAD): the use of computer software to support the design of a product.

Concurrent engineering: carrying out different stages of the design process at the same time.

AQA Examiner's tip

You should be able to give examples of how information technology has affected the product development process.

The impact of modern approaches on the engineering industry

Most competing companies have access to similar materials and manufacturing processes. They may even have similar designs of product. However, how they manage their manufacturing activities will make a big difference to their ability to make a profit.

Originally, skilled craftsmen would make a finished product from the raw materials. Production lines were developed as a method for making larger quantities of products with lower labour cost per part. Each worker on a production line is responsible for one task, which they repeat over and over again. They are expensive to set up but require a relatively unskilled workforce. Each worker only needs to be trained to do one task in the making of the product.

To maximise their potential to make profit, companies can use production lines, and related approaches such as cellular manufacturing, to make products that will be sold all over the world. One aspect of this **globalisation** is that new production facilities are normally placed where they can be operated most cheaply. The operating cost in different countries varies hugely, depending upon local labour rates and **working practices**. For example, in 1999 the average hourly wage of a textile factory worker in Britain was £6.05; in China it was £0.24. Even adding on the shipping costs, for parts with a high labour content it can be more cost effective to make them in a 'lower wage' country.

As it is expensive to set up new manufacturing facilities, a company may decide it is more cost effective to **outsource** some parts or activities. This means that rather than make them, they will buy these from other companies. The '**contractors**' may have better equipment or expertise to make the parts, or be able to make better use of their equipment by working for several different customers.

Objectives

Explain how the management of manufacturing activities can affect profitability.

List some current approaches to the management of manufacturing activities.

Key terms

Globalisation: the process by which companies start operating across the world, rather than locally.

Working practices: the ways in which people work and companies produce parts.

Outsource: to send work to contractors rather than making it within the company.

Contractors: another company paid to carry out an activity or make a part.

Case study

Driving us wild – how car assembly has changed

The first cars were each assembled by skilled craftsmen. A typical car in 1900 cost more than eight years pay for the average worker.

In 1902 the first production line was created. This idea was expanded by Henry Ford when he set up the first Ford car production line in 1914 in Detroit, US, to manufacture the model T, Photo **A**. Cars were placed on a conveyor to move them round the factory. Each worker stayed in the same place and carried out the same single task on each car. The effect of this was to reduce the labour time required to assemble each car from 12 hours 30 minutes to 1 hour 33 minutes. Production line approaches were largely responsible for the large drop in car prices up to 1920, Chart **B**, when a typical car cost about six months pay for the average worker.

A Ford model T

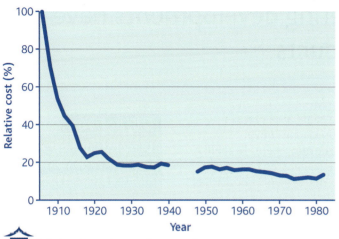

B *Relative change in car prices over time*

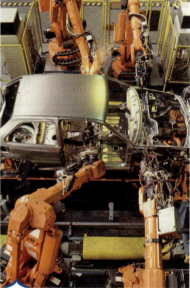

C *Robot assembly on a car production line*

The original Ford factory made every part needed in the car. However, as more factories were opened across Europe to meet the increasing demand, an increasing proportion of parts were outsourced to suppliers. By the 1990s almost all the parts were manufactured by suppliers and Ford, together with other automotive companies, was mainly responsible for assembly. This approach allowed the automotive companies to focus their investment on automated assembly, such as the use of robots, Photo **C**.

The most recent step in outsourcing came in 1996, when Volkswagen opened a new truck and bus plant in Resende, Brazil. This supplies vehicles across South America. The factory comprises six modular areas. Each area builds one part of the vehicle. The novel thing is that these areas are completely outsourced. The labour comes from the contractors that supply the parts. The Volkswagen employees only have to check the finished vehicles at the end of production. This means that Volkswagen only need to directly employ a fraction of the staff needed to build the vehicles.

Activity

Training shoes are a good example of a product with a global demand. Use the internet to find out where they are made and the labour costs in those areas. Compare these with the labour costs in the United Kingdom, USA and some other countries where they might be used.

Summary

Production lines are used to manufacture parts in large quantities at minimum cost.

In the context of increasing globalisation, new production facilities are often placed in countries with low labour costs and favourable working practices.

As it is expensive to set up new manufacturing facilities, companies may decide to outsource some parts or activities to contractors.

AQA Examiner's tip

You should be able to explain how globalisation can affect where a product is made.

How engineering developments have changed products

We have considered how developments in materials, processes, information technology and manufacturing approaches have individually affected engineering and product development. In reality, at any time more than one of these characteristics may be changing. This can be demonstrated by considering how one type of product has changed over time.

Objectives

Demonstrate how a range of different developments can affect a product over time.

Case study

Making waves – the development of radios

The radio comprises a casing and the internal components. It picks up radio waves and converts these into sound. For much of the twentieth century, radios were one of the main sources of news and entertainment for the family. Although the importance of radios has declined due to other technologies (such as televisions, CD players and MP3 players), in 2006 over 15 million new radios were bought in the UK.

The first radio station in the world was established in 1897 on the Isle of Wight, England. At this time radios were built in small quantities by hand, by small companies and craftsmen. The casings were typically wooden cabinets.

In 1889, the first factory to manufacture radios was opened by Marconi in Chelmsford, employing around 50 people. The radios were manufactured by hand. To accommodate the size of the components inside the radio, the casings were the size of a chest of drawers, for example Photo **A**. They cost the equivalent of between one and three months wages for the average worker.

A   *An early radio*

In 1906, engineers at Westinghouse developed a vacuum tube detector, also known as a valve. This improved the performance and allowed the size of the radio to be reduced significantly. Photo **B** shows what the inside of a radio looked like using valve technology.

In the 1920s and 1930s, instead of wood casings many companies started to make the casings using Bakelite, Photo **C**. Bakelite was the first synthetic plastic. This was cheaper and could be formed into complicated shapes more easily than wood.

B   *The inside of a radio using valve technology*

C   *A radio with a Bakelite case*

In 1947, a new type of electrical component was invented: the transistor. This could be used in the radio circuit in place of valve technology, Photo **D**. Transistors are both much smaller and much cheaper than electrical valves. In 1954, a company called Regency introduced the first radio using transistors. This was less than a third of the size of radios which used valve technology.

In the late 1940s, the screw injection moulding machine was invented. This could be used to produce complex shapes with a wide range of plastic materials, more cheaply and more accurately than the existing plastics forming processes. By the late 1950s most radios casings were made using this new technology.

In 1960 Sony launched a pocket-sized transistor radio, which included a plastic casing. The radio was manufactured using a mass production approach in Japan, although the assembly was carried out by hand. Photo **E** shows an example of a mass-produced transistor radio with an injection moulded casing.

D   *Circuit inside a transistor radio*

E   *A mass-produced transistor radio with an injection moulded case*

In 1964, further developments in electrical components allowed a British company, Sinclair, to launch a match-box sized radio. Continuing development over the following 20 years allowed radios to be shrunk further, for example Photo **F**. The controlling factor on the size of the radio ceased to be the components. It became the dials and buttons, which had to be large enough for the user's fingers to operate.

In the 1970s computer-controlled machines were developed which could mount the components in the electrical circuit on the circuit board. This allowed further cost savings and the increased accuracy meant that the size of the circuit could be reduced even further.

In the 1990s, an increasing proportion of radios were manufactured in the Far East, particularly China, to benefit from lower wage rates. Small radios, integrated into other products such as pens, could now be bought for less than £1.

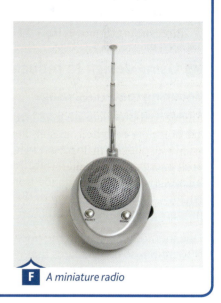

F   *A miniature radio*

Activities

1   In the case study, identify each type of development that occurred: materials, process, information technology, and manufacturing approach.
2   Produce a similar case study explaining how one of the following products has changed over time: wheelchairs, sports shoes/trainers, racing bicycles.

1.8 The impact of the engineering industry on the environment (1)

■ Engineering and the environment

It is fair to say that in the past the engineering industry has contributed significantly to **pollution** and environmental damage. One of the main reasons for this is that we, as a society, placed more importance on our needs for material progress rather than environmental issues. This means that we were more concerned with having larger cars, mobile phones and the comforts we associate with modern life than the environmental impact of making, using and disposing of these products. Engineering companies were more concerned with generating profit than reducing environmental impact. Another reason is that we may not have been fully aware of the environmental impact of our manufacturing activities.

As environmental issues have become more pressing, their relative importance to society has increased. This has placed increasing moral and legal responsibilities on engineering companies. A number of different approaches are being used to minimise the impact of their activities on the environment. These include:

- Modifying the design of products, which is discussed below.
- Recycling materials and components after the end of the products useful life, which is discussed in Topic 1.9.
- Reducing energy requirements to make the products, which is discussed in Topics 1.10, 1.11 and 1.12.

■ Using design to reduce environmental impact

Reducing materials' usage

Engineered products are made from materials and components. These are normally bought from suppliers in standard shapes and sizes. These will be explained in Topic 4.1. The production of these materials uses both energy and raw materials. For example, plastics are made from oil and most metals are made from metal ores, mined from the ground. Further, changing the **form** of these materials into the form of the product requires energy to power the manufacturing processes.

One method of reducing the impact of a product on the environment is to minimise any changes that need to be made to the materials. This reduces the quantity of materials needed. This can be achieved by modifying the design of products so that the materials are used in the standard sizes that are readily available from suppliers. It can also mean removing features that do not contribute directly to how well the product works, such as features added for visual appearance.

Extending product life

Extending product life reduces environmental impact by eliminating the need to make replacement products. This may sound obvious, but it also means that fewer products are sold. This reduces the potential future sales of the manufacturer, which can make some designers reluctant to follow this approach.

Objectives

Outline the relationship between engineered products and environmental needs.

Explain how product designs can be modified to reduce environmental impact.

Information

The 6Rs

The 6Rs are key words representing the different possible approaches to reducing environmental impact.

- Reduce – using less material and energy
- Repair – extend product life rather than disposal
- Recycle – reduce use of resources and energy by reprocessing materials
- Reuse – don't dispose of a product if it could be used to fulfil the same use again or a different use
- Rethink – review if the need is necessary and the features of the design
- Refuse – decline to use products that are not environmentally friendly

Product life can sometimes be extended by using materials with improved properties. For example, rust can be a big problem for car exhaust systems. Corrosion resulting from the combination of heat from the engine and water causes the metal to slowly turn to rust. Eventually, enough metal can turn to rust so that holes form and the exhaust is no longer able to carry out the job it was designed to do, Photo **A**. The use of coatings to improve resistance to rusting has allowed exhaust systems to last for longer before they need to be replaced. Whilst these materials may cost slightly more, the delay in needing to buy a replacement can make them cost effective.

 A rusted car exhaust

For products containing moving parts or electrical components, an effective way of extending life is to design products so that they can be repaired when they go wrong, rather than be replaced. For example, this may involve using standard screws to attach access panels, Photo **B**. This approach also has a disadvantage, in that it can allow people who do not know how to maintain the equipment to tinker with the operation of the product. This is one of the reasons why products such as mobile phones use 'non-standard' screws, which need special screwdrivers.

B   *Removing the access panel to a computer system*

Activity

Carry out a product analysis on a mobile phone. Identify any maintenance activities that might need to be carried out, such as changing batteries or replacing SIM cards, and explain how these have been accommodated in the design. Evaluate whether the phone could be repaired if an electronic component fails and explain your decision.

Key terms

Pollution: contamination of the environment.

Form: the size, shape and condition of a piece of material, product or part.

Summary

In the past, the engineering industry was a significant contributor to pollution and environmental damage. Our society placed more importance on material progress rather than environmental issues. As environmental concerns have become more pressing, their relative importance has increased.

One approach to reduce the environmental impact of engineered products is to rethink the design. This may allow a reduction in the quantity of materials used. It may also allow life extension, through the use of improved materials or design to allow for repair.

AQA Examiner's tip

You should be able to explain how environmental issues affect design decisions during product design.

1.9 The impact of the engineering industry on the environment (2)

When designing products, the best way to minimise the impact on the environment would be to use **sustainable materials**, such as wood, cotton or latex. Unfortunately, the properties of these materials are not suitable for most applications. A compromise is to minimise the amount of new 'non-sustainable' materials used in the product. This can be achieved through design, as explained in Topic 1.8, or through reusing components and recycling materials.

There are requirements in law for reuse and recycling in some industries. For example, The End of Life of Vehicles Directive came into force in Europe in 2000. It gives legal targets for what proportion of a vehicle has to be reused or be suitable for recycling. These targets meant that by 2006, 80% by weight of the materials used in a car had to be suitable to be recycled. This has to increase to 85% by 2015. The responsibility for achieving this is placed on the company that manufactured the vehicle.

Reuse

Reuse means using the component again. This is preferable to recycling, as no energy is needed to change the **form** of the part. For example, scrap yards are a well established source of parts to repair cars, Photo **A**. This is because it is cheaper to buy parts second hand than to buy them new.

Recycling

Recycling means breaking down or melting material and changing it into a different shape, so that it can be used in another application. Recycling applies to both products at the end of their useful life and to waste generated during the manufacture of the product.

Recycling at end of product life

When the average person thinks of recycling, they normally think of piles of old products waiting to be melted down and converted into something else. There is a well established practice of doing this for metal, for example Photo **B**.

Recycling can be carried out for metals, glass and many different types of plastic. Products containing materials that can be recycled are often marked showing the material that they are made from, Table **C** and Table **D**.

It is relatively easy to recycle suitable products that are made from only one material. Many products contain a number of different materials or components which means that they need to be taken apart to allow recycling. Unfortunately, the materials used in some of these products can be very difficult to separate, which makes recycling difficult. For example, composite materials are a very fine mixture of two or more different types of material. They are not normally suitable for recycling. As a second example, it is not normally cost effective to dismantle electronic circuits.

Key terms

Sustainable materials: materials that are grown or produced as a natural product and can be replaced without permanently consuming resources.

Reuse: using the material or component in another application without changing its form.

Form: the size, shape and condition of a piece of material, product or part.

Recycling: breaking or melting down the material so that it can be used in a new form.

A *Car bumpers in a scrapyard awaiting reuse*

B *Crushed cars waiting to be melted down*

Process waste and by-products

Most engineering processes produce some waste or by-products. For example, the material removed by machining operations, the smoke produced by welding, and the old oil that is replaced in engines when they are serviced.

Materials removed by machining operations can be collected and sent for recycling.

Dust and gases normally need to be removed from the working environment for safety reasons. This can be achieved through fume extraction. Dust particles are often a mix of different elements, which makes them difficult to recycle. For this reason they are often disposed in landfill. This means that they are buried along with other refuse and rubbish.

Some liquids can be recycled. For example, contaminants can be removed from water used in metal plating by a process called 'ion exchange'. Other liquids may be suitable for reuse. For example, waste oil from frying food can be collected to power engines and generators. However, a proportion of other industrial liquids still need to be stored in sealed containers and sent for landfill.

D *Meanings of recycling symbols*

♻	Product contains plastic materials that can be recycled
(alu)	Product is made from aluminium and can be recycled
△	Product is made from glass and can be recycled
↻	Packaging contains some recycled materials

C *Meanings of the numbers in the plastic recycling symbols*

Symbol	Type of Plastic	Typical Uses
1 PETE	Polyethylene terephthalate	Drinks bottles, oven-ready meal trays
2 HDPE	High density polyethylene	Washing-up liquid bottles
3 V	Polyvinyl chloride	Cling film, drinks bottles
4 LDPE	Low density polyethylene	Carrier bags, bin liners
5 PP	Polypropylene	Margerine tubs, microwavable utensils
6 PS	Polystyrene	Yoghurt pots, foam meat trays, packaging
7 OTHER	Any other plastics that do not fall into the above categories	

Summary

When it is not possible to use sustainable materials, the impact on the environment can be reduced by minimising the amount of new 'non-sustainable' materials used.

If components can be reused, no energy is needed to change the form of the part.

Recycling involves changing the form of a material, so that it can be used in another application. Not all materials and components can be recycled.

The materials used in a product might be recycled at the end of the product's usable life. In some cases, waste and by-products produced during manufacture can also be recycled or reused.

Activity

Collect examples of plastic packaging showing each of the different recycling numbers.

AQA Examiner's tip

You should be able to explain how solids, liquids and dust/gases need to be controlled, disposed of or reprocessed.

1.10 The impact of the engineering industry on the environment (3)

Energy use in manufacturing

Every manufacturing activity in an engineering company is dependent upon energy. Most of this energy will go directly into manufacturing the product. For example:

- The manufacturing processes may be carried out on machines powered by electric motors.
- Heat treatment processes may use gas or electricity.
- The lorries used to **transport** raw materials and take away finished products normally use diesel fuel, made from oil.
- Additionally, energy will be needed to run some operations of the factory that are not directly involved in manufacturing. For example, the lighting and computers will be powered by electricity. The factory and offices may be heated using gas or electricity.

The cost of the energy used by a company can be significant. The current trend is that these costs are increasing. If a company can reduce what it pays for its energy, then this may give it a significant advantage over its competitors.

How industry obtains its energy

In the past, the main considerations when selecting an energy source have normally been availability and cost. For most processes, the type of energy needed is determined by the nature of the **process**. For example, electric machines need to run on electricity! However, heat treatment, such as operating an oven to fire clay pots, can be carried out using either gas or electricity. The decision on which energy source to use will normally be based on the amount of gas or electricity needed, taking into account the efficiency of the process. However, a company may also consider the environmental implications. The electricity may have been generated using gas, in which case much more gas would have been used, due to process inefficiencies and power losses during transmission. This could mean that, overall, less resource would be used if gas was used to heat the oven directly.

For most companies, electricity can be bought through a provider who takes it from the national grid. The national grid is the high-voltage electric power transmission network that connects power stations with all the areas of the UK. A visible feature of this network in rural areas is the network of electricity pylons, Photo **A**. Similarly, gas can be purchased from a gas supplier.

For companies who use a lot of energy, it may be cost effective to generate their own power. This means that the costs of setting up the power generation equipment and of maintaining and running the equipment would need to be less than their normal power bills. This is most often done by companies that melt down metals, such as steel makers, or large chemical plants.

Objectives

Outline how energy is used during manufacturing.

Explain how a suitable energy source is selected for an engineering application.

Describe how industry can dispose of, or use, surplus energy by-products.

Key terms

Transport: the movement of materials and goods.

Process: an operation that changes the size, shape or condition of a material.

Heat: thermal energy.

A Electricity pylon

Use and disposal of surplus energy by-products

There are a number of steps that a company can take to reduce the amount of energy that it uses. Examples of these are shown in the 'Ways to Save Energy' text box.

It may also be possible to 'reuse' some **heat** energy which would otherwise be allowed to escape into the environment through cooling stacks and towers, Photo **B**. For example, in the chemical industry, chemicals are often heated so that they react together to form a product. The product is then allowed to cool down. If the energy released during the cooling-down phase could be transferred to the next chemicals instead, this would reduce the amount of new energy needed to heat them up. This can be achieved by using a 'heat exchanger'. This is a device built to transfer heat energy from one medium to another. One of the more common designs of heat exchanger is a condenser, Photo **C**. In addition to chemical plants, heat exchangers are used in heat treatment processes, refrigeration systems, air conditioning systems and power generation plants.

B *Cooling towers releasing steam in a power plant*

C *Condenser section of a heat exchanger at a manufacturing plant*

Activity

Using the internet, identify the activities carried out to manufacture one of the following products: the steel frame of a bicycle, a clay vase, or a foil-wrapped chocolate egg. This should include any transportation of raw materials and shipping to the customer. Identify the type of energy used at each step in the process.

Information

Ways to save energy

- Power down and turn off electrical equipment when it is not in use. This includes machines, computers, and lights.
- Replace old machines with more energy efficient new machines.
- Insulate the facilities used to heat-treat products. This can reduce the energy needed to heat them by 20%.
- Insulating the factory can reduce general heating bills, just like in domestic homes.

Summary

Every manufacturing activity in an engineering company uses some form of energy, most typically electricity, gas or diesel fuel.

The main considerations for an engineering company when selecting an energy source are normally cost and availability.

Surplus heat energy is often allowed to escape into the environment. However, it can sometimes be reused through the use of heat exchangers.

AQA Examiner's tip

You should be able to explain the decisions made in industry when deciding on using one process or energy source rather than another, e.g. gas versus electricity for heating to high temperatures.

1.11 Using technology to reduce environmental impact (1)

World energy consumption has been growing each year, Graph **A**. If we continue to meet this increasing demand using traditional approaches, we will create significant **pollution** and damage to the environment. One of the most important ways that technology is being used to reduce environmental impact is through the development of **renewable** methods of generating electricity. To understand these, we first need a general understanding of how electricity is generated.

Generating electricity

There are three main approaches used to generate electricity: energy conversion using electric generators, photovoltaic cells or chemical reactions.

Electric motors are used in a wide range of applications, from CD (compact disc) players to toothbrushes to cars. Supplying electricity to the motor makes it spin. Electricity can be generated by using this process in reverse, to create a generator. Electricity is produced by turning the motor in a generator using a turbine, which itself is turned by an external force. The turbines can be very large, Photo **B**. The different methods of generating electricity by this approach are distinguished by the type of force that spins the turbine. This is the most widely used approach to generating electricity.

Photovoltaic cells convert light into electricity. Small cells are used to power calculators and garden lights. The larger the area of the cell the more electricity it produces.

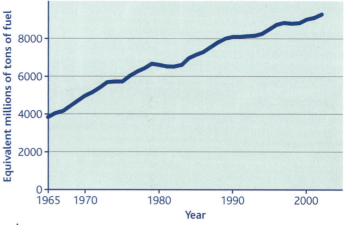

A *World energy consumption*

Chemical reactions can produce electricity. This approach is the principle of battery design, where electricity is generated by a controlled, slow reaction over time. This is not a **sustainable** process, as the chemicals change as a result. However, in rechargeable batteries the reaction is reversed during the period of charging. Unfortunately, some of the chemicals used in batteries can have adverse environmental effects at the end of their usable life. Whilst this approach is widely used for energy storage, it is rarely used as a method of power generation.

Different sources of energy

Sources of energy can be broadly grouped into two types: **non-renewable** and renewable. Non-renewable sources are consumed and will, at some stage, run out. Renewable sources are naturally replenished; these will be discussed in Topic 1.12.

■ Non-renewable energy sources

Fossil fuels

Traditionally, most electricity has been generated using **fossil fuels** such as oil, petrol, diesel, gas and coal. It has been estimated that in 2006 fossil fuels were used to generate 86% of the global electricity supply.

The fossil fuels are burned in large furnaces. The resulting heat is used to create steam which powers the turbines in a generator. This is a proven and reliable technology. However, the by-products of using fossil fuels cause significant environmental damage. This includes pollution, acid rain, smog and gases which contribute to global warming. Fossil fuels are also a non-renewable resource. Our supplies will eventually run out. This also means that as supplies of fossil fuels reduce and demand increases, the price of fossil fuels is likely to increase.

B *Turbines used to generate power in a hydroelectric plant*

Nuclear power

Nuclear power uses radioactive materials. As a result of the radioactive nature of this material, when enough of it is put together in a 'nuclear pile' it heats up. This heat is used to turn water to steam, which is used to turn the power turbines, Diagram **C**.

Nuclear power plants can produce electricity at a low cost. However, it is very expensive to build new plants. A lot of attention has to be given to safety, as uncontrolled radioactive materials can cause significant damage to health and the environment. Nuclear power plants also create radioactive 'nuclear waste'. Some of this waste is radioactive material that was been used in the pile. The other waste is materials that have been exposed to the radioactivity.

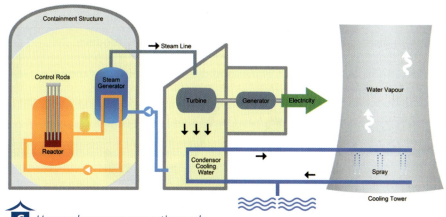

C *How nuclear power generation works*

Summary

The most common used method of generating electricity is by energy conversion using an electric generator driven by a turbine. Electricity can also be generated by photovoltaic cells and chemical reactions.

Renewable sources of energy are naturally replenished. Non-renewable sources of energy are consumed when used.

The use of fossil fuels to generate electricity is a proven technology. However, they are a non-renewable resource. Their use creates pollution and contributes to global warming.

Nuclear power produces low-cost electricity. However, it is very expensive to build nuclear power plants.

Activity

Create a list of all the different items of electrical equipment that you have used in the last week. Against each write down what alternatives you would have to use if electricity was not available.

AQA Examiner's tip

You should be able to explain the differences between renewable and non-renewable energy sources.

1.12 Using technology to reduce environmental impact (2)

Renewable energy sources

Wind power

Electricity can be generated by using the wind to turn a turbine in the form of a propeller. The rate of turning is normally quite slow, so the propeller shaft is normally connected to a gearbox, as shown in Illustration **A**. The gearbox increases the rate at which the rotor of the generator is turned, allowing it to generate electricity even in quite low winds. A large wind turbine can have propeller blades of ten metres or longer, Photo **B**. A 'wind farm' of six or more large wind turbines can produce enough electricity to meet the needs of several thousand homes. It is also possible to use small wind turbines with propeller blades of 30 centimetres or less, to supply electricity for home use or to power small equipment.

The main disadvantage of wind turbines is that they only produce energy when it is windy. Further, some people think that it is not attractive to have large wind turbines and dislike the noise that they make when working.

B   *A large wind power generator*

A   *Cutaway view showing the inside of a large wind power generator*

Tidal energy

Energy can be produced by the movement of the tide at sea using similar principles to wind power. However, here the turbine blade is located under water, Illustration **C**. This costs more to set-up than wind power, as it is more difficult to put into position.

Biofuel and biomass

Biofuel and biomass are organic substances. Biofuels may be vegetable oils or liquids which have been fermented and contain alcohol. Biomass includes solid and fibrous materials such as plants and wood, as well as animal manure. Similar to fossil fuels, these materials can be burnt to power generators. In general, weight for weight they are less efficient than fossil fuels at generating energy. Although they are a **renewable** resource, the by-products of this method include gases which contribute to global warming.

C   *How tidal power is used to generate energy*

Geothermal energy

Similar to nuclear energy, geothermal energy also uses steam to power the turbines. In this case, the water is heated by molten rock under the surface of the Earth. The places where this method can be used are limited to where the hot material is within a few miles of the surface, such as Iceland. This is because it would be very difficult to drill the holes, from which the water and steam is released, to much greater depths. Where they can be built, geothermal plants are much cheaper than nuclear plants.

Hydroelectric power

Hydroelectric power uses water pressure to generate electricity. The water pressure is created by building a dam across a large river, for example Photo **D**. This creates a large, artificial lake. The water escapes from the lake by turning the turbine. This can be very expensive to set up. The artificial lakes can be very large, which has a significant effect on the local environment. It was estimated that by February 2007, between 40 and 80 million people had been forced to move their homes, as the area where they used to live was now covered by lakes created by hydroelectric dams. In addition to the land used, hydroelectric dams also have a significant impact on marine life.

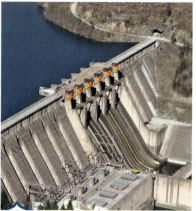

D   *A dam used to generate hydroelectric power*

Solar energy

Generating electricity from solar energy differs from the other forms in that it uses **photovoltaic** cells. The amount of electric current generated is related to the area of cells used. A 'solar energy farm', such as the one shown in Photo **E**, may cover the area of several sports fields. This approach can easily be used in homes and factories, by positioning solar panels on the roof. Whilst these may not fully satisfy all of the energy needs of the building they cover, they would make a significant contribution. The main disadvantage of solar energy is that maximum electricity is only generated when it is sunny.

E   *Solar panels*

Activity

Create a table summarising the advantages and disadvantages of each of the renewable and non-renewable methods of generating electricity. This should include the equipment required, the costs of setting up and operating the equipment and any environmental implications. Use the internet to find information where it is not included above.

Key terms

Renewable: replenished through natural sources.

Photovoltaic: generates electric current when exposed to light.

Summary

Wind power generates electricity by using wind to turn a turbine.

Tidal energy uses the movement of the tide to drive a turbine, generating electricity.

Biofuel and biomass can be burnt to power generators.

Geothermal energy generates electricity by using steam heated by molten rock under the surface of the Earth to power the turbines.

Hydroelectric power uses the pressure of water passing through a dam to turn a turbine, generating electricity.

Solar power generates electricity using photovoltaic cells.

AQA Examiner's tip

You should be able to compare different sources and their relative cost/availability and environmental factors.

1

Engineered products are developed to meet the needs of users. How effectively these needs are met depends upon a range of factors, including the materials that the product can be made from and the processes used to design and manufacture it. Developments in these factors can have a significant impact on engineered products and society.

The development of new materials with improved properties, or new combinations of properties, can lead to the significant improvements in the performance of a product or even to new products and markets. Improvements to existing materials may occur several times over the life of a product. This may allow products to meet existing needs better or to meet additional needs. These improvements could include improved properties or changes in the form of the materials being used. Either new or improved materials may allow changes in the design of a product.

Improvements in the processes used to make products can help to reduce constraints due to process capability or manufacturing cost. These improvements are often achieved through the application of computer control.

The use of information technology can also lead to advantages during the other stages of the product development process. In particular, the use of Computer Aided Design to create drawings and models has significantly changed the role of the designer. Generally, the use of information technology has led to reduced product costs and allowed products to move from being an initial idea to a finished item much faster.

The use of information technology has also made information transfer easier, both within companies and to contractors when outsourcing. This has helped to support increasing globalisation, with new production facilities often being located in countries with low labour costs and favourable working practices.

Over a period of time, the needs and priorities of users can change. As a society, the importance that we place on environmental issues has increased. Industry is responsible to minimise the impact of processes used in manufacturing on the environment. This can involve reducing waste and recycling by-products. It can also mean reducing energy requirements or obtaining energy from sustainable sources.

∞ links

Further information on CNC machines
www.technologystudent.com/cam/camex.htm

Further information on CAD
www.cadinschools.org/index.php.

Further information on environmental legislation
www.envirowise.gov.uk/page.aspx?o=MBEN4PBHSM

Further information on the environmental effects of energy generation
www.niauk.org/images/stories/pdfs/energy-in-environment.pdf and
www.greenpeace.org.uk/files/pdfs/nuclear/nuclear_economics_report.pdf

Further information on nuclear power
http://science.howstuffworks.com/nuclear-power.htm

AQA Examination-style questions

1 Give two examples of how developments in materials have affected the design of radios. *(4 marks)*

2 Explain, with an example, how the development of a new material can lead to a new product. *(4 marks)*

3 State two process constraints that a designer may have to take account of when developing the design for a new product. *(2 marks)*

4 One reason for the development of new processes is to manufacture new materials. State one new manufacturing process and the material it was developed to be used with. *(2 marks)*

5 Identify two ways in which new technology is used in the design and manufacture of televisions. *(4 marks)*

6 A company specialises in tuning-up and servicing high performance sports cars. They sometimes need to modify engine parts or make replacements. Explain why they may decide to outsource the machining rather than carrying out the manufacture themselves. *(4 marks)*

7 Complete the following table to show how the application of information technology has changed the product development process. *(4 marks)*

Activity	How it was carried out …	Could now be carried out …
Creating drawings	Hand-drawn by draughtsmen	
Testing ideas	Making prototypes and physical models	
Placing orders for materials	Telephone and post	
Assembling the product	By human workers	

8 The manufacture of sports shoes used in the UK is often carried out by contractors in India and China. Explain one advantage and one disadvantage of this outsourcing. *(4 marks)*

Double award

9 Explain how the development of electronic components has affected the design of mobile phones. *(3 marks)*

10 Give two advantages of using a CNC milling machine to make a flat surface when manufacturing a batch of components. *(4 marks)*

11 Explain how the use of information technology can be used to manufacture products of different specification on the same production line. *(4 marks)*

12 Explain how the development of information technology has affected the role of the engineering designer. *(4 marks)*

13 Explain the waste disposal issues caused by the use of plastics for manufactured products. *(4 marks)*

14 A company carries out contract machining services. List three ways that the company can reduce its use of energy. *(3 marks)*

15 Explain what is meant by 'sustainable energy'. *(2 marks)*

16 List one advantage and one disadvantage of using wind power to generate electricity. *(2 marks)*

17 Describe how damage to the environment will be reduced by applying one aspect of modern technology. *(4 marks)*

Objectives

Explain the steps of the design process.

Explain why, in practice, the design process is often not a simple sequential activity.

The main driving force in the development and production of an engineered product is normally to make a profit. Companies look for opportunities where customers have a need that they could meet. They then consider developing a product (or service) that could satisfy this need. This may involve months of investigation, to carefully calculate:

- the possible demand – how many of the new product they will sell
- the estimated cost of producing the product
- the likely profit to be made.

If this initial study shows that the development of a new product is justified, the company will ask a designer or design team to develop it.

The designer must make sure that the product meets all the needs of the customers, whilst making the best possible profit for the company. This can be a very challenging task, requiring many different things to be considered. In this chapter, we will investigate how a product is designed and how this design is communicated between engineers. To do this, we first need to understand the steps that need to be carried out when designing a product.

A *The design process*

The design process

The manufacturing company writes a design brief, which is a short statement of what is required. This is given to the designer or design team, who carry out research and analysis to determine exactly what is needed in detail. This might include academic research, but normally it includes things like finding out every detail about what the customer wants and how it could be made.

The results of these investigations are used to create a list of needs that the product must satisfy. This will include customer needs, such as what the product will look like, product needs, such as how the product will work, and manufacturing needs, such as what processes must be used to make the product.

The designer will then create a number of different possible solutions, making sure that they meet all the needs in the specification. They, or the company, will select one or more of the best ideas and develop them further to test if they will work in practice, using drawings, computer simulations and physical models. They will also prepare drawings to communicate these ideas to other people involved in the development and making of the product.

The designers and engineers will then work out how the design could be made. They will probably start by making a prototype, to check that it can be made and will work. The prototype will be tested against the design brief and the needs in the specification, to make sure that it meets the needs of the customer and the company. As a result of this testing, the design or manufacturing processes may be changed, or they may decide that they can proceed to make the product.

The design process in practice

In practice, the design process is rarely carried out as a simple sequence of tasks. Often things are found out as you go through the process that change what was needed at an earlier step. For example:

- The design brief might have to be changed following feedback from potential users during the research.
- The ideas developed may need to be changed when planning the making of the product, as it might not be possible to make the design to the accuracy needed.

These might mean that the designer has to jump back to earlier steps several times during the development of the product.

Starter activity

A company has decided to develop a new mobile phone. For each step of the design process, identify at least one result or finding that could mean that they would need to go back to an earlier step in the design process.

AQA Examiner's tip

Don't skip stages of the design process or jump directly to a solution. If you do, it will probably mean that your solution will not be as good as it could be.

Record your thinking, i.e. why you made decisions, at each stage of the process in case you need to make changes later.

Using a client design brief

What is a design brief?

Once a company has identified a need, they will prepare a **design brief**. This is normally a short statement, often no more than one paragraph. The purpose of this statement is for the company or **client** to outline to the designer what they require. Ideally, the design brief should include:

- A statement of the need to be met or the problem to be solved – i.e. the **function** that the product must carry out.
- Who the end user will be and who would buy the product, if these two are different – the potential market.
- The important product features or **user needs**.
- Any **constraints** – these are the things that limit what you can make.
- Diagram **A** shows the example of a brief for an emergency wind power generator.

The design brief should not state the detail of the solution or outcome. The more open that the brief is, the more opportunities it gives the designer to come up with a novel or profitable solution.

Objectives

Explain what a design brief is and the information it should include.

Explain how to analyse a design brief and identify other design needs.

The design brief is to produce a wind power generator for use in emergency situations. It should be able to provide enough power to recharge a mobile phone. It should be suitable for use by campers aged 16 and upwards and must be able to fit inside a backpack. It should be safe to use and suitable for production in batches of six using standard workshop equipment.

Statement of problem to be solved

Identifying the user

Important features

Manufacturing constraints

A Design brief for an emergency wind power generator

Analysing the design brief

The first thing to do when analysing the design brief is to list the key features and constraints outlined above. The designer will then try to identify any other design needs not shown in the design brief. There are many different types of possible need which he should think about. This analysis of the needs might be in the form of a spider diagram or a table listing the questions to be investigated, such as Table **B**. For a simple product, there may be just 20–30

B Questions that might be asked to identify possible design needs

Type of need	Example of a question that may be asked
Aesthetic	What colour is the product? What finish or texture does the surface have?
Cost	How much should the product cost?
Customer	Who is the product designed for? When, where and how will they use it?
Environment	Can the product be recycled? What power source does it use?
Safety	What safety standards does the product need to meet? What must be done to make sure that it is safe for the user?
Size	How big should the product be?
Function	What is it intended to do? How could it do this?
Maintenance	Do you have to be able to change batteries or take it apart to repair it?
Materials	What properties are needed from the materials that will be used to make the design?
Manufacturing	Are there certain processes that must be used to make the product?

questions. When doing the detailed design of a complex product with many components, such as a car, there may be thousands of questions to answer to identify all of the design needs.

Research and analysis

There are a lot of different ways of finding out the answers to the questions identified during the analysis of the design brief. Sometimes a single piece of research may answer more than one of the questions. Types of research carried out might include:

- Market research – identifying what similar products are available and their features.
- Product analysis – looking at or taking apart examples of products that address similar needs.
- Carrying out tests to determine how the product should function.
- Looking up sizes in tables of ergonomic data – this means seeing what sizes the product should be so that the user is able to use it easily. For example, you might design an MP3 player so that it can be handled easily by a 7-year-old. This would be a different size to one designed for the hand of a 20-year-old.
- Carrying out questionnaires and interviews to determine user needs.

The designers will normally include comments on each piece of research showing what this means for the design, such as: 'As a result of this, the design should … '. This is because sometimes different types of needs will contradict each other, which will be explained in more detail in the next topic. If that happens they will need to know where each need has come from, so that they can decide which one is the more important.

C *Using a mobile phone while in a remote place – but how will it be recharged?*

Activities

1 Choose an electrical product that is used in the home. Write the brief that you think was given to the designer for that product.

2 Write separate design briefs for the following two cars: a small hatchback for use in a large city and a sporty hatchback. Explain the differences between these two briefs and what you think that these might mean for the design of the product. For one of these cars, create a list of the other information that the designer would need to know to be able to design a successful product.

Summary

The design brief is a short statement of the need that must be met, written by the client for the designer.

The design brief should include features that are important to the design, such as constraints and user needs.

The designer analyses the design brief to start creating a list of needs that the product must meet. The analysis may also create a list of other questions that need to be answered to be able to design an effective product.

Key terms

Design brief: a short statement of what is required.

Client: the company or person for whom the work or design is being carried out.

Function: what the product is intended to do.

User needs: the things that the customers require the product to do.

Constraints: things that limit what you can make.

AQA Examiner's tip

Use a real, external client on any assessed work and pay regard to any feedback or advice.

What is a specification?

The design specification is a list of all the needs that the design must meet. Most of these needs are answers to the questions that were identified when analysing the brief and investigated during the research. It should include, as a minimum, the design **constraints** and user wants, relating to the **form** and **function** of the product. Table A is a design specification for the emergency wind power generator which was explained in the previous topic.

The designers will normally include comments justifying why each of the needs is important. They may also list where that particular need was identified during the research. There are two reasons for doing this:

- Some needs may have to be a compromise. For example, for the wind turbine, the designers may have found technical information showing that they would need the blades of the turbine to be as large as possible to generate lots of power. They may also have investigated a range of back packs, to work out the size and form of blades that could be carried. They would either have to make a decision as to which piece of research was the more important, or to find a compromise where the design manages to satisfy both of the requirements.

- It can also be very important where, due to other issues later in the design process, it is necessary to go back and change a design.

Why is the specification important?

The designers will use the specification to determine the **design parameters** that they have to work to. This means that the specification limits what the designers can do. If any relevant needs have been missed from the specification, the designers may design a product that does not do what potential customers or manufacturers may need it to do.

A *Example of a specification for the emergency wind power generator*

No.	Need
1	The wind turbine shall generate at least 100 milliamps at 12 volts in a wind speed of 5 metres per second.
2	The wind turbine shall be robust enough to operate in a wind speed of up to 5 metres per second for at least one hour.
3	The wind turbine shall be mounted on a rod, which is between 7 mm and 9 mm in diameter.
4	The mounting rod shall have a length of at least 80 mm.
5	The blades shall rotate within an operating envelope of less than 400 mm^2.
6	The wind turbine shall rotate within an operating envelope of less than 350 mm.
7	The wind turbine shall include a tail, to position the blades into the wind, which shall have a surface area of at least 20 000 mm^2.
8	The wind turbine shall have 6 blades, each with a minimum surface area of 4800 mm^2.
9	The blades and tail shall be made from a material which can be recycled.
10	The blades and tail shall be yellow so that they are easily visible from a distance of 5 m.
11	The body of the shell shall be made of a material which is resistant to corrosion in rain water.
12	There shall be no sharp edges on either the blades or the body of the wind turbine.
13	The cost of the parts for the wind turbine shall be less than £20.00.
14	The total cost of the wind turbine including labour shall be less than £100.00.

Using the specification to evaluate the product

Design ideas and the final design proposal should be tested against the specification, to see if they meet all of the needs. For this reason, it is important to make sure that every need is **quantifiable** and can be objectively tested. One way of doing this is to make sure that all of the listed needs have five characteristics, known by the acronym SMART:

- They must be Specific. For example, for the wind turbine, the designer would not say 'it needs to work well', as different people may have different understanding of what this means. The designer could say 'it shall generate at least 100 milliamps at 12 volts', as all engineers would understand this.

- They must be Measurable. For example, the designer would not say that the blades of the wind turbine need to be big, as different people will interpret 'big' differently. The designer could say 'the minimum surface area of the fins shall be 4800 mm^2'.

- They must be Achievable. A common mistake is to use an exact measurement, such as 'the mounting rod shall be 8 mm diameter'. This would mean that if the mounting rod was 7.999 mm or 8.001 mm diameter it would be a failure! The designer would state either a range, such as 7 mm to 9 mm in the case of the mounting rod, or for other criteria a maximum or minimum value.

- They must be Realistic. There would be no point in writing down a need that could never be achieved!

- They must be Time-bound. For example, there can be a big difference between two products where one needs to work for one minute and the other for one year.

It is important that the specification is not limited just to the **functional requirements**. It has an important role in being used to evaluate that both operational activities and the product cost meet the requirements.

Summary

The design specification is a list of all of the important needs that the product must meet.

To ensure that the design proposal and final product can be tested against the specification, the needs should be specific, measurable, achievable, realistic and time-bound.

Key terms

Constraints: things that limit what you can make.

Form: the size, shape and condition of a piece of material, product or part.

Function: what the product is intended to do.

Design parameters: the values for characteristics that the design has to satisfy.

Quantifiable: measurable.

Functional requirements: things that are needed for the product to meet the customers' needs.

AQA Examiner's tip

Ensure specifications generate testable and quantifiable outcomes.

Generating and developing design ideas

After the needs that the product must meet have been established, the next step is to design a product that can meet these needs. This involves generating a range of ideas and then developing some of them into a design proposal. There can be a lot of bouncing between ideas generation and ideas development – after the first ideas are captured, they may be developed and improved, leading to more new ideas.

The designs start in the mind of the designers. They can then use a number of different methods to capture, **visualise** and communicate these ideas.

Objectives

Explain the different uses of drawing and sketching.

Explain how modelling can be carried out to help with the development of designs.

Sketching and drawing

Sketches and drawings provide a visual representation of a design.

Sketches and drawings do different jobs. The aim of **sketching** is to communicate an idea. Traditionally, most designs start by being captured on paper as sketches. The aim of a **drawing** is to communicate detailed information about a design, such as dimensions, what parts to use and how the parts should be put together.

Drawings are one of the main methods used by designers and engineers to communicate information about products. There are a wide variety of different types of drawing used in engineering. Most drawings have to conform to certain rules, known as standards or **conventions**, so that any engineer can understand them.

Sketching and drawing will be covered in detail in later topics.

Case study

Development of vacuum cleaners at Dyson

A *Sketching initial ideas*

B *CAD modelling*

C *Making a physical model from card*

Once the needs have been established, a designer sketches some initial ideas, Photo **A**. These will be evaluated to see how well they could meet the needs in the specification. Successful ideas will be drawn using CAD. This will allow the fit of parts to be modelled, Photo **B**. It could also be used for computer-based simulations of performance. A physical model will be produced using card, Photo **C**, so that the size and shape can be evaluated, to see how attractive they would be to a user. Once the designers are happy with the fit of the parts and the appearance, a working prototype will be made. This will be tested and evaluated against the needs in the specification, Photo **D**.

D *Testing prototypes*

Modelling

Models provide a 3-D representation of a product. Most designers use **modelling** to help them develop and test out their ideas. The models made can be virtual, where they only exist on a computer, or physical, where a part is made.

Virtual modelling can be carried out using CAD software. For example, these can test whether parts will fit together or whether electronic circuits will work before a product is made. This can save time and avoid expensive mistakes.

Physical models are constructed before making a product as it is much quicker and cheaper to correct mistakes and make improvements at the design stage than when a product is in manufacture. Physical modelling approaches such as card modelling and rapid prototyping are used to check the fit or overall appearance of an item. These approaches use materials that are different from those that would normally be used in the final product. For example, the materials used during rapid prototyping range from polymers to flour. Approaches such as constructing electronic circuits without using printed circuit boards (bread-boarding) allow the function of designs to be tested.

The final type of physical modelling used for most products is prototyping. This means making a one-off, or a small batch of the product. Prototypes are used to **evaluate** how well a design works and the manufacturing processes that could be used to make it.

Key terms

Visualise: create an image of what a design will look like.

Sketching: a quickly produced visual representation of an idea.

Drawing: a visual representation of an idea to communicate detailed information.

Conventions: rules of presentation that drawings must conform to.

Modelling: making a three-dimensional representation of a product.

Evaluate: compare how well a product meets the design needs.

Activities

1. Write a magazine article explaining the process of product development at Dyson and how this benefits both the customer and the company.

2. Automotive companies are continually redesigning the cars that they make. Explain how they could use a number of different approaches to modelling during the design of a new car.

AQA Examiner's tip

Explore as many possibilities as you can in the early stages of designing.

Draw on research to establish a good foundation for ideas.

Only use research that is clearly relevant to the product.

Summary

Sketching is a method of capturing and communicating ideas.

Drawing is used to communicate detailed information about design ideas. Drawings must normally conform to conventions, so that they can be understood by any engineer.

Modelling provides a 3-D representation of a design that can be used to test and evaluate it. They can be used to check how well parts fit together or how well a design works.

2.4 Sketching

What is sketching?

The aim of **sketching** is to **communicate ideas**. Sketches often have a large number of labels to explain or point out interesting features.

Sketches are often used to get initial ideas down on paper, for example as shown in Diagram **A**. This is sometimes known as capturing ideas or producing concept drawings. Freehand sketching doesn't need any drawing equipment apart from a pencil or pen. The sketch should be produced quite quickly. However, this doesn't mean that it is rushed or unclear.

Unlike most other types of drawing used in engineering, there are no standard rules or conventions for producing a sketch. The standard of sketching varies a lot between different designers and different companies. They do not have to be produced to **scale**. Sketches can be 3-D or 2-D. 3-D sketches are often used to show the whole object. 2-D sketches are often used to show close-up views of individual features or details.

Objectives

Explain how sketching is used.

Explain how to render a sketch, using different line thicknesses and shading.

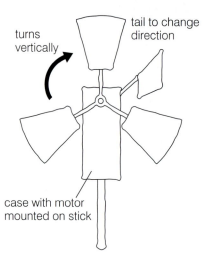

turns horizontally — 2 twisted blades — turns vertically — tail to change direction — generator in here — case with motor mounted on stick

A *Examples of sketches: ideas generation for an emergency wind power generator*

Rendering

Rendering means adding colour or texture to a picture. Although sketches are produced quickly, they are often rendered to give a more realistic view of what the product could look like.

Two common forms of rendering are 'thick and thin lines' and 'shading'.

Thick and thin lines

Different line thicknesses can be used to make parts of illustrations stand out. A simple, easy to use technique is:

1 The sketch is produced as normal, using thin lines to show where any two surfaces meet, Diagram **B a**.

2 On any edge where only one surface is seen, the line is then increased to medium thickness, Diagram **B b**.

3 On any outside edge, the line is made even thicker, Diagram **B c**.

This is easy to do if sketching is carried out in pen, as technical pens are available in different sizes, such as 0.25 mm, 0.5 mm and 1 mm. It can also be done in pencil, but care is needed to keep the line thickness consistent.

Shading

If an object is placed near a window or light, the side facing the window will appear to be a lighter colour than the side which is in shade. These lighter and darker versions of the same colour are called tones.

An object with one tone all over looks flat and uninteresting. A sketch shaded with different tones looks more realistic. With round objects, like balls or pipes, it also helps to make the surface appear curved.

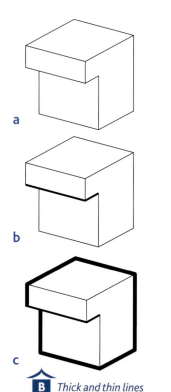

a

b

c

B *Thick and thin lines*

Shading can be easily carried out using a pencil. For example:

1   Lightly shade the whole object, Diagram **C a**.

2   For any sides or areas that are further from the light source or only receive light indirectly, shade them again, so that they become a bit darker, Diagram **C b**.

3   For any sides that have no direct exposure to the light source, shade them again, so that they become even darker, Diagram **C c**.

Shading is normally most effective if a single colour is used, with different tones. Using too many different colours can reduce its impact.

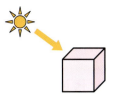

a Begin by lightly shading the whole shape

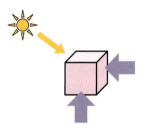

b Now shade sides or areas further away from the light source again

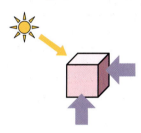

c And finally shade sides not directly exposed to the light source again

C   *Shading*

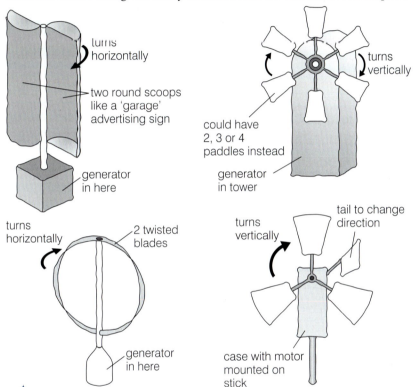

D   *Examples of rendered sketches: ideas generation for an emergency wind power generator*

Activities

1   Produce a series of quick sketches of a house. On different versions of the sketch, use 'thick and thin lines' and 'shading' to see how these make it look more realistic.

2   Produce a series of quick sketches of a sports car. Use both 2-D and 3-D. Render your sketches to make them look more realistic.

Summary

Sketching is a method of communicating ideas.

There are no conventions for producing a sketch.

Sketches and drawings can be made to look more realistic by rendering. Rendering techniques include controlling line width and shading.

Key terms

Sketching: a quickly produced visual representation of an idea.

Communicate ideas: share a concept with others.

Scale: the ratio of the size of the design in the drawing to the size of the part or finished product.

Rendering: applying colour or texture to a sketch or drawing.

Shading: create different tones on a sketch or drawing.

AQA Examiner's tip

Master a few simple rendering techniques, such as controlling line width and shading, to make your sketches look more realistic.

2.5 | Isometric drawing

▮ What is isometric drawing?

Isometric drawing is a method of creating a 3-D view of a design idea or a product. Although it doesn't give a perfectly accurate representation of the object, isometric drawings look more like the final product than drawings produced by many other techniques.

The main features of an isometric drawing are:

- ▪ The closest part of the object to the person looking at it is a corner.
- ▪ All horizontal lines are at an angle of 30° to the horizontal, Diagram **A**.
- ▪ All the vertical lines remain vertical.
- ▪ All measurements are to **scale**.

Special paper for isometric drawing is available to provide guidelines for the drawing, as used in Diagrams **A** and **B**.

▮ How to produce an isometric drawing

One way to produce an isometric drawing is to start with the front corner, then gradually extend this line out to become the full image. For example, the following steps can be used to draw a simple cube, Diagram **B**.

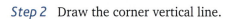

Step 1 Look at the object you want to draw from a corner.

Step 2 Draw the corner vertical line.

Step 3 Add any horizontal lines extending from the corner line, at 30° to the horizontal.

Step 4 Add any vertical lines extending from the horizontals.

Step 5 Add any horizontal lines extending from these verticals and so on, repeating steps 4 and 5 as needed.

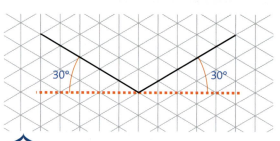

A *The angle used for isometric drawing, shown on an isometric grid*

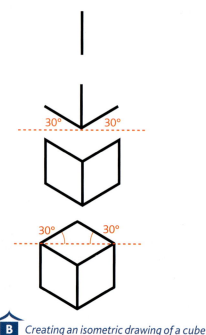

B *Creating an isometric drawing of a cube*

This approach is ideal for parts where every edge is straight, but designers also need to know how to draw parts with curved features. Fortunately, if someone can draw an isometric cube, almost any other shapes can be created using a technique called **crating**.

Crating

When using isometric projection, curved shapes, such as circles and arcs, cannot be drawn with compasses. They have to be constructed so that they are in the correct view, Diagram **C**. One way of drawing them is to use crating, to provide guidelines. Another way, if one is available, is to use an isometric template.

To create a circle, first an isometric box, called a crate, is drawn. Ideally, the lines for the crate should be very faint. It is then marked where the circle will touch the edges and these points are joined up, Diagram **D**. After drawing, the lines for the crate should be erased. Shapes with curved features are drawn in the same way.

The rendering techniques outlined in the section on sketching can be used to further improve the presentation of isometric drawings. Diagram **E** shows an isometric drawing of one idea for the emergency wind power generator.

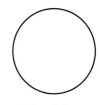

Isometric view   'Direct' view

   Isometric view of a circle compared with a 'direct' view of a circle

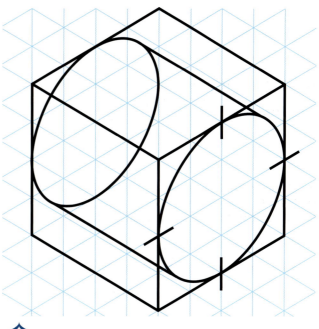

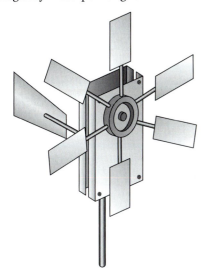

E   Rendered isometric drawing of an idea for an emergency wind power generator

D   Drawing an isometric cylinder

Activities

1. Produce an isometric drawing of a pirate's treasure chest. Don't forget to include the lock and some bands of metal running around the chest, for reinforcement.

2. Produce an isometric drawing of a sports car. Render this image to make it look even more realistic.

Summary

Isometric drawing is a method of creating 3-D views of an object.

On an isometric drawing, the horizontal features of the product are shown at 30° to the horizontal. Vertical features remain vertical.

Isometric crates can be used to give guidelines for circular or curved shapes.

Communicating designs through orthographic drawings

Why are orthographic drawings important?

To make a part, an engineer needs to know the dimensions of every feature on it. Sketches and isometric drawings normally do not contain enough information on the sizes. Orthographic projection is a formal drawing technique which is used to communicate the dimensions of a part.

Orthographic projection is the most widely used method of producing technical drawings for engineering products. The drawings produced are called **orthographic drawings** or working drawings. It is vital that any designer, engineer or machine operator can look at a working drawing and understand it. For this reason, there are a large number of rules and **conventions** that specify how the drawing and the information it contains must be presented. These rules will be covered in the next section.

To machine operators and manufacturing personnel, these are the most important drawings. They use them to find the dimensions when making the parts, Photo **A**. If any of the dimensions on the drawing are wrong, or any of the features are missing, then the manufactured product will not be correct. This means that it is vital to be accurate and pay attention to detail when creating working drawings.

What is orthographic projection?

An orthographic drawing usually shows three 2-D views of a part. There are several different possible combinations of views. However, the most common form of orthographic drawing is called 'third angle projection'. This is represented by the symbol shown in Diagram **B**. A third angle projection shows the part from the top, which is called the plan elevation, the side, which is called the end elevation, and the front, called the front elevation. On the drawing, the three elevations are laid out in an L-shape. The front view is underneath the plan view, and the sides of these two views are exactly in line. The end view is to the right of the front view, and the top and bottom of these views is exactly in line, Diagram **C**.

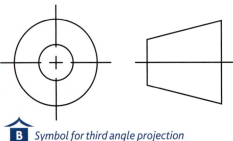

B *Symbol for third angle projection*

For comparison, Diagram **D** shows the layout of an orthographic drawing using first angle projection. The three faces in a first angle projection are on the opposite sides of the part to those shown in third angle projection.

A *Using the working drawing to check parts during manufacture*

Information

Blueprints

Working drawings are sometimes also called 'blueprints'. This term dates back to when all drawings were done by hand. As the drawing was produced, there was a sheet of blue carbon paper underneath it that the design also appeared on. This 'blueprint' of the design was used on special equipment to print copies of the drawing.

Preparing to use orthographic projection

Before computers became widely available, every drawing was produced by hand. Companies employed teams of skilled draughtsmen to turn their designers' ideas into working drawings. The drawings were produced using black ink and drawing boards. Whole rooms were dedicated to storing the drawings produced.

To draw an orthographic projection of a design by hand, it is useful to have the following equipment:

- drawing board with a T-square
- pencils (2B and HB) and eraser, or fine liners
- rule
- set square.

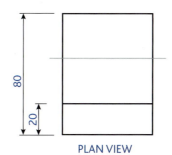

PLAN VIEW

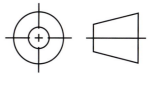

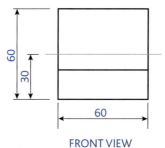

FRONT VIEW

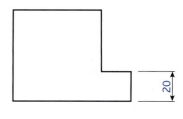

END VIEW

C *Third angle projection*

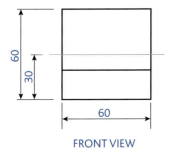

FRONT VIEW

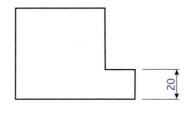

END VIEW

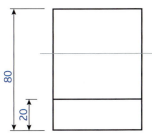

PLAN VIEW

D *First angle projection*

Key terms

Orthographic drawing: a working drawing of a part showing three views, to communicate the dimensions of the design.

Conventions: rules of presentation that drawings must conform to.

AQA Examiner's tip

Ensure that your orthographic drawings are laid out properly, showing the correct views.

Summary

The purpose of an orthographic drawing, also called a working drawing, is to communicate the dimensions of a part.

A third angle orthographic projection includes the plan, end and front views of the part.

Activities

1 Using the internet, find a range of examples of orthographic drawings using third angle projection.

2 Produce an isometric drawing of the component shown in Diagram **C**. It must be to the correct scale.

Standards and conventions

An orthographic drawing must be drawn in a certain way. This is to make sure that anyone using an orthographic drawing can understand it. British Standards BS308, PD7308 and PP7308 give the **conventions** and the symbols that should be used to represent different features.

Objectives

List the standards and conventions used in orthographic drawing.

General presentation

It is a general requirement that the drawing should have a border and a title block, Diagram **A**. The border normally contains zone references, so that people discussing the drawing can easily be directed to the features that they are talking about. The title block should contain:

- The drawing name or title.
- The drawing number.
- The scale used for the drawing.
- The name and signature of the person who made the drawing.
- The drawing issue number and its release date.
- The projection used (normally shown as the symbol).
- Any other information needed to make the part. For example, this could include the material to be used.

Planning the drawing

Working drawings need some consideration and planning before the part is drawn.

Firstly, the part must be drawn to **scale**. This means that its size must be in proportion to the size of the finished part. It should be stated as a ratio. The recommended scales are:

- 1:1 (full size)
- 1:2, 1:5, 1:10, 1:20, 1:50, 1:100, 1:200, 1:500, 1:1000 (smaller than full size)
- 2:1, 5:1, 10:1, 20:1, 50:1, 100:1 (larger than full size).

A scale of 1:5, means that if the side of a component was 200 mm long, the line on the drawing representing this side would be 40 mm long. The scale chosen should allow all of the features of the drawing to be clearly visible, whilst still fitting on the paper.

Drawing the part

When drawing the part, there are specified styles of line that have to be used for different types of feature, Diagram **B**.

- Geometry lines show the visible edges or features of the part. These should have solid lines.

SCHOOL OR COLLEGE NAME GOES HERE

DRAWN:	DATE:	TITLE:	
CHECKED:	DATE:		
APPROVED:	DATE:		
REVISED:	DATE:		
MATERIAL:		DRAWING NUMBER:	
FINISH:		SCALE:	SHEET OF

A   *Layout of a typical formal drawing*

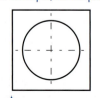
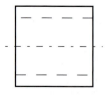

——————— Geometry line

– – – – – Hidden line

– · – · – · – Centre line

For example, for a square part with a hole through it:

B   *Line styles*

- Hidden lines show features beneath the surface of the part, such as holes. These should have dashed lines.
- Centre lines show the centre of any circular features in the part, such as holes or the centre of a shaft. These should have lines comprising alternate long and short dashes.
- Leader lines (see below) have faint solid lines.

Showing the dimensions

Dimensions showing lengths on working drawings should always be in millimetres. Angular dimensions are normally in degrees.

Leader lines project from the part to indicate what points the dimension runs from and to. These should not touch the feature on the part itself, Diagram **D**. The dimensions are written above the dimension line. For lines measuring dimensions in the vertical direction, this means that they are on the left of the line. The unit of measurement is not normally shown, as this is specified in the title block.

Diagram **C** shows a working drawing of one of the components designed for the emergency wind power generator discussed earlier in the chapter, including all of the standards and conventions.

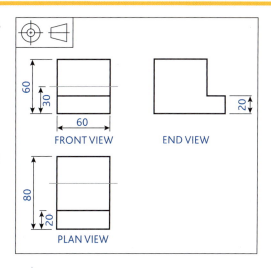

C   *An orthographic drawing for one of the components in the emergency wind power generator*

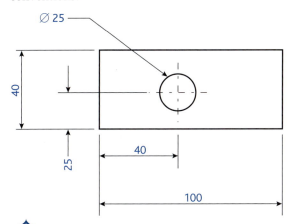

D   *Examples of dimensioning*

Activities

1. On Diagram **C** identify the different features required by convention for the general presentation and the different line types.

2. Create a dimensioned isometric drawing showing the component in Diagram **C**.

Summary

British Standards BS308, PD7308 and PP7308 give the conventions that must be used in an orthographic drawing.

There are detailed conventions covering the layout of the drawing and the types of lines used to draw and dimension the part.

Dimensions for any lengths should be in millimetres.

Key terms

Conventions: rules of presentation that drawings must conform to.

Scale: the ratio of the size of the design in the drawing to the size of the part or finished product.

AQA Examiner's tip

Ensure that your drawing and dimensioning meets all the standard conventions.

Computer Aided Design (CAD) is the use of software packages to assist in the design of a product. When most people talk about CAD they are thinking of producing high quality drawings. However, CAD also includes the use of software to model and design electrical, pneumatic and mechanical systems. These will be investigated in Topic 2.10.

Objectives

Explain what is meant by CAD.

Explain the differences between 2-D and 3-D CAD for drawing.

Outline the main uses of CAD drawings.

The use of CAD for drawing

CAD approaches have now largely replaced manual drawing for creating working drawings. The reasons for this will be covered in the next section. It is used all over the world, to develop products ranging from simple nuts and bolts to televisions, cars and houses.

There are two main types of CAD drawing software. One type is for drawing in two dimensions. The other type is for drawing in three dimensions. Both types of software normally include a wide range of drawing tools. These are the basic commands used to create CAD drawings. The tools typically range from drawing simple lines and inserting shapes, through to duplicating and manipulating the drawn features and modifying and deleting features.

A *Creating a 2-D CAD drawing*

2-D CAD

Drawing using 2-D CAD software is similar to drawing by hand. The screen has a working area, which is in effect the piece of paper that you draw on. This area can be changed to almost any size, and you can zoom in to see features close up, Photo **A**.

3-D CAD

Using 3-D CAD, a three-dimensional model of the part being drawn, is created, Illustration **B**. Starting to draw a 3-D shape is more complicated than drawing a 2-D shape. In effect, you have an empty three-dimensional space that you will fill with the design. You first have to create an imaginary piece of 2-D paper within this space. This is often called a plane or **workplane**. The design starts as a 2-D drawing on this workplane, which is often a **sketch**. Once this sketch has been created, it can be extruded to give it thickness or depth. Other **features** can then be added to this design. Alternately, existing features can be changed by

B *Creating a 3-D CAD model*

editing this design. In addition to being able to change the work area and zoom in or out, the design can also be rotated and moved on the screen, so that it can be viewed and edited from any direction.

Most 3-D CAD software also has the ability to produce working drawings directly from the CAD model.

Uses of CAD drawings

The main use of CAD drawings is to provide the dimensions that are used to make the product, either as working drawings or by the electronic transfer of data to computer-controlled machines. The data in the CAD software can also be used for **modelling** the design and to carry out computer simulations of the testing of the design, for example Illustration **C**.

C *Virtual testing of a component in the Dyson vacuum cleaner*

Activities

1 Using the CAD software available in your school, produce a drawing of a lunar buggy that could be used to drive two astronauts around on the surface of the Moon.

2 Using the internet, identify a number of different types of product that were designed using CAD. You should find pictures of the original CAD designs and the finished products and explain any differences between them.

Summary

Computer Aided Design (CAD) is the use of software to assist in the design of a product.

The use of 2-D CAD drawing software generates drawings in the same manner as drawing by hand.

3-D CAD drawing software can be used to create both 3-D models of designs and 2-D drawings.

Although the main use of CAD software is to produce working drawings, it can also be used for modelling and simulated testing of the designs.

Key terms

Computer Aided Design (CAD): the use of computer software to support the design of a product.

Workplane: in 3-D CAD, the features are first drawn on a 2-D surface.

Sketch: in 3-D CAD, a 2-D drawing produced on a workplane.

Features: details on the design.

Modelling: making a three-dimensional representation of a product.

AQA Examiner's tip

Use the technology to allow you to demonstrate your ideas and, if possible, to output the information to a computerised machine.

Developing design ideas using CAD for drawing (2)

The advantages of using CAD for drawing

The use of **Computer Aided Design (CAD)** software has become widespread because it has many advantages over traditional approaches. Compared with drawings produced by hand:

- It can be quicker to create a new drawing, as **features** can be copied from previous designs and edited. This saves money by reducing labour costs.

- It is easier and quicker to make changes to a drawing. To make changes to a drawing made by hand, it is often necessary to restart the drawing from scratch. To make changes to a CAD drawing, you can open and edit the existing file.

- CAD drawings can be more accurate.

- CAD drawings can be saved electronically. This saves lots of space and the associated costs (such as rent). Before CAD was used, companies had whole rooms full of the drawings used to make their products. All of this data can often now be stored on a single computer or a few DVDs.

- CAD drawings can be easily circulated to anyone who needs them, using CDs, e-mail or downloads from the internet.

- Standard parts, such as screws, nuts and bolts, can be downloaded from libraries of CAD parts. This means that they don't have to be drawn, saving lots of time.

- It requires less manual skills in graphics to produce high quality drawings.

- The CAD drawings can be sent electronically to computer-controlled machines that will be used to make the parts. Some CAD software can also work out the machining instructions for the machines, saving time for the operator setting them up or programming them.

CAD models produced using 3-D packages can have additional advantages:

- The software can often automatically generate working drawings that satisfy the standards and **conventions**.

- The designers can use the software to model how well the parts would fit together in the finished products. This is done by putting together drawings of the different parts that are used in the assembled product, Illustration **A**. It means that they can see how well the parts fit together, so if there are any problems, they can change the design before any parts are made. This saves cost and time.

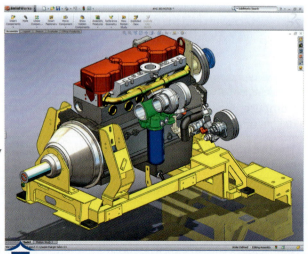

A *Creating a 3-D model of a part to check the fit of components*

The images can be rendered, to provide an accurate representation of what the finished part or assembly will look like, Photo **B**. It is possible to quickly change the appearance of any CAD parts, comparing what the product would look like if it was made in different colours or using different materials.

The models of the product can be tested virtually. For example, stress analysis can be carried out by the computer to see if a product would be strong enough, Photo **C**; or, for a car, the design could be tested in a simulated wind tunnel, showing how aerodynamic it is. This allows changes to be made to the design without the cost or time of making a product to test.

B *A rendered image of a product assembly*

C *Example of a stress analysis: the different colours represent different stress levels*

The disadvantages of using CAD for drawing

CAD approaches can also have some disadvantages compared with drawings produced by hand:

- It is harder to keep the drawings safe from competitors, as the electronic files can be easily copied.
- The CAD software can be very expensive. In general, 2-D CAD software is less expensive than 3-D CAD software.
- Specialist training is normally needed to operate the CAD software.
- Work can be interrupted or destroyed by computer viruses, corrupt files and power cuts. For this reason, regular back-ups should be taken of all important files.
- Fewer people are needed to carry out drawing work, so fewer draughtsmen are employed.

Activities

1. Using the internet, investigate the range of different tests that can be carried out using virtual CAD models. Create a table with three columns, listing the type of product, the type of testing carried out, and explaining how the results of this testing can affect the design of the product.

2. 'Rapid prototyping' uses CAD drawings and models to make physical parts that can be used to evaluate different characteristics of the design. Using the internet, identify at least four different methods of 'rapid prototyping', noting down the advantages and disadvantages of each.

Key terms

Computer Aided Design (CAD): the use of computer software to support the design of a product.

Features: details on the design.

Conventions: rules of presentation that drawings must conform to.

Summary

The main advantage of using CAD software for drawing is that drawings can be produced more quickly, accurately and cheaply than manual drawing.

3-D CAD models can also allow the fit and function of the design to be evaluated virtually, saving the cost and time of making prototypes for testing.

The disadvantages of using CAD software for drawing can include the cost of the software and the increased risk that drawings may be copied for use by competitors.

AQA Examiner's tip

You should be able to describe at least two different examples of the use of CAD drawings and explain the advantages of CAD in these applications.

2.10 Developing design ideas using CAD modelling

Most people associate **Computer Aided Design (CAD)** with producing high quality working drawings. However, CAD software can also be used to carry out other tasks during the development of **electrical**, **electronic**, **pneumatic** and **mechanical** products. In these applications, CAD software is available which can select suitable components for the working parts of the product and then model the function of the finished product.

Objectives

Explain that CAD can be used to design and model the functional parts of engineering systems

Using CAD to design engineering systems

Many engineered products contain working parts. These parts may be stationary, such as electrical circuits, or able to move, such as pneumatic circuits or mechanical systems. A single product can contain a large number of different parts. It is the combination of all these parts working together that decides how well the product functions.
For example:

■ The electronic circuit used in a basic mobile phone may contain over 500 components, including 18 or more computer chips, Photo **A**.

■ A pneumatic circuit used for safety switches on a computer-controlled machine may contain 10 or more components.

■ A vehicle gear box may contain 10 or more gears, Photo **B**.

A *An electronic circuit board from a mobile phone*

Component selection

When designing engineered products, selecting the correct combinations of parts is critical if the product is to perform as needed. This normally involves carrying out a large number of mathematical calculations to work out which individual components should be used.

For many applications, CAD software can calculate which components to use. The designer will tell the software what the product is required to do and enter the **operating parameters**. The software will then carry out all of the calculations required and recommend suitable components. This can save a considerable amount of the designer's time, therefore reducing costs. For example:

■ The CAD software used to design electrical and electronic circuits can calculate which components should be used, how these components need to be arranged and even provide the component layout. Diagram **C** includes an example of a circuit diagram and the design of a printed circuit board (PCB) that could be used to build the circuit, both generated using CAD software. Some of the software packages used to do this can also be directly linked to computer-controlled machines that can manufacture the PCBs.

B *Internal view of a gearbox*

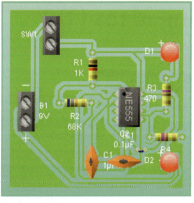

Real-world view

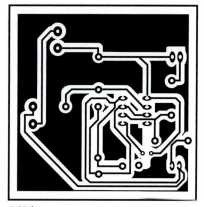

PCB layout

 C *Modelling of a simple alarm circuit*

- For mechanical systems, the gear ratios and sizes needed, along with characteristics such as the number of teeth per gear and the tooth profile, can be calculated.

Modelling performance

Many of the CAD software packages also allow the system to be modelled and tested virtually. This saves a lot of time and cost, by reducing the number of prototypes that need to be made. For example, Diagram **D** shows the simulated testing of a pneumatic circuit.

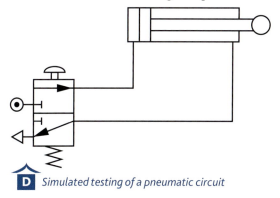

D *Simulated testing of a pneumatic circuit*

Key terms

Computer Aided Design (CAD): the use of computer software to support the design of a product.

Electrical: circuits containing simple conductors that allow current to flow through them.

Electronic: circuits including semi-conductor materials, such as computer chips.

Pneumatic: powered by air.

Mechanical: moving parts, or parts for use in moving equipment or machinery.

Operating parameters: the conditions within which the product must operate.

Activities

1. Using the CAD software for electrical modelling available in your school, create an alarm circuit that could be used to protect a house against break-ins.

2. Write a magazine article explaining how the end customer benefits from the use of CAD during the design of an electrical product of your choice. You should use a maximum of 200 words.

Summary

CAD software can be used to calculate which components to use in products containing electrical, electronic, pneumatic and mechanical parts.

It can also be used to test how well virtual models of these products work, saving the time and cost of making a prototype.

AQA Examiner's tip

You should be able to clearly explain at least two different examples of the use of CAD. Use CAD software to model and develop any electrical or pneumatic systems in your design

2.11 Other ways of communicating ideas (1)

Assembly drawings

Assembly drawings are also called general assembly drawings. They are mainly used for complex products containing many parts, showing the complete finished product with all the parts assembled in the correct place. They get their name because they are used by fitters when they are assembling the parts to make the product.

They are presented like working drawings, Diagram **A**. They may include overall dimensions, instructions on how to put the product together and information on the surface finish requirements. They normally also include a list of the parts to be used, with each part clearly identified on the drawing.

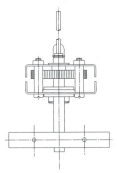
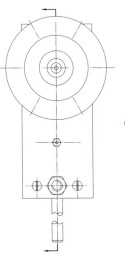
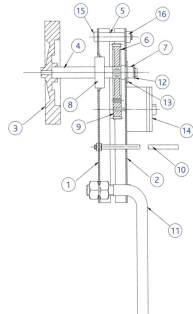

ITEM	DESCIPTION	No OFF	MATERIAL
1	FRONT PLATE	1	ALUMINIUM
2	BACK PLATE	1	ALUMINIUM
3	HUB	1	PLASTIC
4	MAIN SPINDLE	1	BRASS
5	SPACE	4	PLASTIC
6	LARGE GEAR	1	PLASTIC
7	BUSH	1	BRASS
8	BALL BEARING	1	STEEL
9	SMALL GEAR	1	PLASTIC
10	TAIL ROD	1	STEEL
11	MOUNTING ROD	1	STEEL
12	SPIRE	1	STEEL
13	CLIP	1	STEEL
14	SOLAR DC MOTOR	1	
15	SCREW	4	STEEL
16	CLINCH NUT	4	STEEL

THIRD ANGLE PROJECTION	DRAWN		ALL DIMENSIONS IN mm	
	DATE 28/07/06	SCALE	WIND TURBINE	ASSEMBLY DRAWING
	CHECKED	1:1		
	DATE			SHEET 3 OF 10

A *An assembly drawing for a wind turbine*

Exploded views

An **exploded view** is a picture of the product that is taken apart. The individual components are drawn separately, but in the same order as when they are assembled. This gives a good idea of how the product is put together. They have many uses:

- Maintenance fitters use them when taking the product apart to service or repair it.
- Personnel who have not been trained how to read orthographic drawings may use them when assembling simple products.
- Sales and marketing staff use them when producing promotional literature.

Both exploded views and assembly drawings can be produced very easily if there is a 3-D CAD model of the product, Figure **B**.

Schematic diagrams

Schematic drawings show how the working parts of a product are related to each other. They use standard symbols for the individual parts, rather than the drawings of the parts themselves. This means that any engineer should be able to understand them.

The most common form of schematic diagram is an electrical (or electronic) **circuit diagram**, for example Topic 2.10 Diagram **C**. The symbols used conform to standards such as British Standard PP307. These will be covered in detail in Chapter 3.

A schematic diagram of an electrical circuit is more useful than a picture. It contains additional information, such as component values. However, someone making the circuit may find it difficult to relate the schematic to what the circuit actually looks like. For this reason, in addition to the circuit diagram, a wiring diagram is often produced, showing where individual wires and components go on the circuit board.

Another type of schematic diagram can be used for pneumatic and hydraulic systems. An example of a schematic drawing for a pneumatic system, generated using CAD software, was shown earlier in Topic 2.10, Diagram **D**.

a

b

c

B *Creating an exploded view of a valve from a 3-D CAD model*

Summary

Assembly drawings are working drawings that show the complete finished product with all the parts assembled in the correct place. They are often used by the fitters assembling the product.

An exploded view is a picture of the product that is taken apart. They are often used for maintenance activities.

Schematic diagrams use standard symbols to show how the working parts of a product are related to each other. They are commonly used for electrical, electronic, pneumatic and hydraulic systems.

Key terms

Assembly drawing: a working drawing showing an assembled product, with all the parts shown.

Exploded view: a drawing of a disassembled product showing the components in the correct relationship to each other.

Circuit diagram: a schematic diagram using symbols to show how electrical or electronic parts relate to each other in a product.

AQA Examiner's tip

Produce accurate drawings which show your ability to work across sectors and use conventions and symbols.

2.12 Other ways of communicating ideas (2)

Systems diagrams

A **system** is a collection of interconnected parts that exists to perform a function. Any product that is an assembly of parts and interacts with its environment in some way is a system.

A **systems diagram** is a representation of how a system will work. It breaks down the functions to be carried out into simple categories and lists the parts needed to carry out that category. This means that if someone can create a systems diagram, they can produce an overview of the functional parts needed in the design of any product that is a system.

What are the parts of a system?

Diagram **A** shows the parts of a simple system. The power supply (for example, the battery, mains electricity or compressed air) is never shown as part of the system.

The signal

A very common mistake when creating a systems diagram is to overlook the signal. Every box must be connected by a signal. The signal shows the route into, through and out of the system. The signal must be measurable. However, the type of signal can be different between boxes. Common signals include movement, electricity, light and sound.

Objectives

Explain what a system is.

List the parts of a systems diagram.

Explain how a systems diagram is used.

Information

Examples of systems

Cars

Engines

Vending machines

Mobile phones

MP3 players

Computers

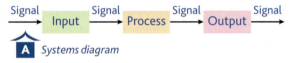

A *Systems diagram*

The process boxes

Every process box in the system must have a function. This means that they must change or process the signal in some way. For example, they could change the size of the signal, such as increasing or reducing the electrical current. Alternatively, they could change the type of signal, such as converting electricity into light or air pressure into movement.

Each of the process boxes is normally a physical item or an assembly of physical items. A simple system may have just one process box of each type, for example Diagram **B**. Complicated systems may have several process boxes, for example Diagram **C**.

B *Example of a systems diagram for a drinks machine*

Input

The input is the thing that starts the system. This is normally some form of sensor. For example, the inputs in a simple house alarm may include a keypad for entering the alarm code, a contact switch on a door and a motion sensor.

Process

The process is the 'brain' of the system. It says what should happen when the system is activated. In the case of the alarm, this may include a latching function so that when a sensor is activated the alarm stays on and a timer function is activated. This will stop the siren after five minutes, to avoid upsetting the neighbours. These could both be achieved with a single electronic circuit, containing a component called a **microcontroller**.

Output

The output is how the system responds to being activated. For example, for the alarm, it may be a siren and a flashing light. The output will convert the electrical signal from the process box to sound and light signals respectively.

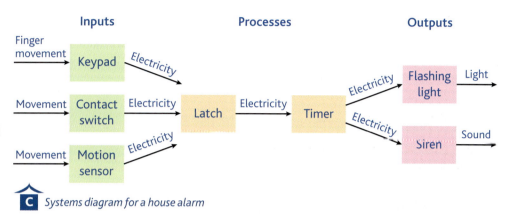

C Systems diagram for a house alarm

Designing a robot buggy part 1

The brief was to design a simple, low cost robot buggy. Once activated by the user, it had to be able to move along the floor until it hits an object, then turn away at a right angle to the object.

After researching and analysing the problem, the systems diagram shown in Diagram **D** was developed. The buggy would be controlled using a microcontroller.

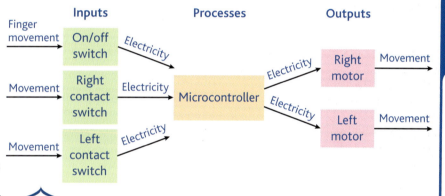

D Systems diagram for a robot buggy

It would have an on/off switch and two contact switches – one on each side, so it would know which way to turn. The movement and change of direction could be achieved by having two central wheels, each operated by a separate motor.

Case study

Activities

1. Produce a systems diagram for a security light that sits outside a garage. It should be activated when an intruder comes near and stay on for 15 minutes. Clearly label the signal, input, process and output.

2. Produce a systems diagram for a laptop computer.

Summary

A system is a collection of parts that exists to form a function.

The parts of a system are signals, inputs, processes and outputs. A complex system may have several process boxes of each type.

A systems diagram is a useful tool to help identify the functional parts that need to be included in a product.

Key terms

System: a collection of parts that interacts with its environment and performs a function.

Systems diagram: a schematic representation of a system.

Microcontroller: a computer-on-a-chip, containing a processor, memory, and input/output functions.

AQA Examiner's tip

Use a systems diagram to work out the parts that need to be included in your design.

Flowcharts

A **flowchart** shows the order in which a series of events are carried out. A common mistake when designing a new product is to confuse flow diagrams with systems diagrams. They show different things. A **systems diagram** shows the functional parts of a system – the processes used to carry out the tasks. A flowchart shows the instructions that control what the system does. In this type of application, the flowchart could be thought of as providing the 'intelligence' or program for the process box in a systems diagram.

Objectives

Explain the difference between a flowchart and a systems diagram.

Describe how to draw a flowchart.

Describe how flowcharts are used.

Drawing flowcharts

Different symbols are used to represent the different possible types of event – some of these are shown in Diagram **A**.

- A terminator is used for the start and end of the series of events.
- The process box is used when an instruction or action must be carried out.
- An input/output is used where information or items are added or given out.
- A decision represents a question or choice. It must have a yes or no answer.

The complete set of symbols and their meanings are listed in British Standard BS4058. The event or question is written inside the symbol.

The symbols are linked together by arrows which show the correct sequence of events. The arrows should always be either horizontal or vertical, never slanted.

When drawing flowcharts, the blocks and symbols should be kept a uniform size. If they are available, stencils can help in doing this.

Terminator

Process

Input/Output

No

Yes

or

Direction arrows

Decision, showing direction arrows

A *Some of the symbols used in a flowchart*

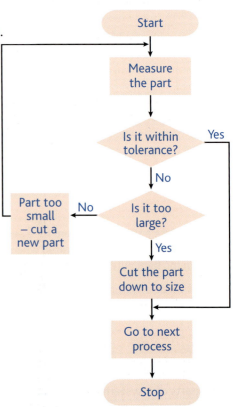

B *A simple flowchart used for quality control during the cutting of a steel sheet*

Use of flowcharts

One common use of flowcharts in manufacturing is as part of the instructions for quality control, Diagram **B**. After a measurement is carried out, a decision is needed. If the product is correct, then it will proceed to the next manufacturing process. If it is incorrect, there may be further questions to determine why it is wrong, so that an appropriate action can be carried out to fix the problem. After the problem is corrected, the part will need to be remeasured, returning to the start of the process. This creates what is called a feedback loop.

Another common use for flowcharts is to create the programmes for computer-controlled systems, such as microcontrollers, robots and computer-operated machines. An example of this is shown in the case study below.

Flowcharts can also be used to plan the sequence of actions during the manufacture of a product. This can be carried out as part of the development of the production plan, which will be covered in Topic 4.10.

Key terms

Flowchart: a diagram showing a sequence of operations.

Systems diagram: a schematic representation of a system.

AQA Examiner's tip

Use a flowchart to plan a sequence of events, such as the tasks you will need to carry out to make your product.

Designing a robot buggy part 2

Case study

As outlined in Topic 2.12, the brief was to design a simple, low cost robot buggy. Once activated by the user, it had to be able to move along the floor until it hit an object, then turn away at a right angle to the object. In the systems diagram, it was determined that it would have two contact switches and two motors.

To develop the control programme, the first step was to write out the instructions.

When turned on, the robot buggy would move forward. When it touches a wall with the right sensor it stops, reverses a short way then turns left and moves forward again until it touches another wall. When it touches a wall with the left sensor it stops, reverses a short way then turns right and moves forward again until it touches another wall. This was then used to develop a flowchart, Diagram **C**.

Flowchart to control a robot buggy **C**

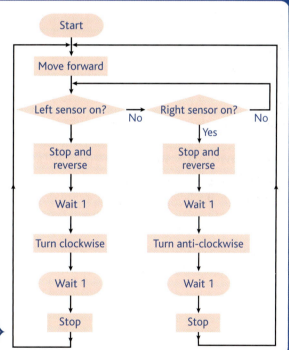

Activities

1. Produce a flowchart to control the sequence of operations at a set of traffic lights.

2. Produce a flowchart to control the automatic doors at the entrance to a supermarket. The doors have a sensor that will detect when someone approaches them and should open automatically. They should close automatically when the person has passed through them.

Summary

A flowchart shows a sequence of operations or events.

Each event in the sequence is shown inside a symbol. Each symbol has a specific meaning. The symbols are joined up by arrows, which show the direction of the sequence of tasks.

Flowcharts can be used to provide instructions for manufacturing tasks such as quality control, to programme computer-controlled systems and to sequence tasks, such as the manufacturing activities for an engineered product.

2.14 Presenting the proposal to the client

The need for approval by the client

As a result of the research, development and any modelling carried out, the designer should be confident that the design proposal can meet all of the **client's** requirements. However, proceeding to production could involve significant costs for materials and new manufacturing tools and equipment. The final decision on whether the product should proceed to production needs to be made by the client.

The final design proposal is normally presented to the client for approval. The client needs to be given confidence that the product meets the brief and the identified needs. He or she will probably be presented with a range of different **media** to see what the product would look like and how it would work, such as:

- presentation drawings of the proposed product, working drawings and exploded views
- models or prototypes of the product
- systems diagrams, circuit diagrams and flowcharts
- images (or videos) of products of similar design or prototype products in use, with tables of technical data and test results

When selecting which media should be presented, the technical knowledge of the client must be considered.

Comparing the design proposal to the specification

The client must be given confidence that the design proposal will meet the needs of the application. One method to do this is to include in the media presented to the client a detailed comparison of the design proposal to the needs in the **specification**, as shown in Table **A**.

If a model or prototype has been made during the design process, then this can be tested and compared with each need. It is likely that a number of different tests will have to be carried out. Where possible, the testing should be objective. This means that it should be based on facts and numbers, rather than subjective opinions.

Client feedback

Having considered the design proposal, if the client is not confident in the proposed designs abilities to meet the needs of the application, or if he thinks that further changes may allow the finished product to better meet the needs of the market, the client may give **feedback** to the designer. He may request changes to the design proposal before proceeding. It is also possible that during the evaluation the client might identify additional needs which he feels need to be considered. This might mean changes or additions to the specification. This could, in turn, mean having to return to an earlier step in the design process, such as selecting an alternative idea for development or further developing the design proposal.

It is essential that any changes requested by the client are considered by the designer. The client is paying for the design either directly or though his wages, so he has the right to expect that his requirements will be met. It is important, therefore, that there is good **communication** between the client and designer.

A *Example of an evaluation against the specification for the design proposal for the emergency wind power generator*

No.	Need	How evaluated	Could the design meet this need?
1	The wind turbine shall be able to generate at least 100 milliamps at 12 volts in a wind speed of 5 metres per second.	Used an ammeter/voltmeter to test output from the motor when modelling the circuit	Yes
2	The wind turbine shall be robust enough to operate in a wind speed of up to 5 metres per second for at least one hour.	By looking at similar existing designs	Yes
3	The wind turbine shall be mounted on a rod, which is between 7 mm and 9 mm in diameter.	By checking that material of suitable size is available and including this in the working drawing	Yes
4	The mounting rod shall have a length of at least 80 mm.	By including this in the working drawing	Yes
5	The blades shall rotate within an operating envelope of less than 400 mm.	Checked the dimensions on a 3-D CAD model	Yes
6	The wind turbine shall rotate within an operating envelope of less than 350 mm.	Checked the dimensions on a 3-D CAD model	Yes
7	The wind turbine shall include a tail to position the blades into the wind, which shall have a surface area of at least 20 000 mm^2.	Calculation based on measurements in the working drawing	Yes
8	The wind turbine should have 6 blades, each with a minimum surface area of 4800 mm^2.	Calculation based on measurements in the working drawing	Yes
9	The blades and tail shall be made from a material which can be recycled.	By looking at the materials used in similar designs	Yes
10	The blades and tail shall be yellow so that they are easily visible from a distance of 5 m.	By looking at similar designs	Yes
11	The body of the shell shall be made of a material which is resistant to corrosion in rain water.	By looking at similar designs and a materials handbook	Yes
12	There shall be no sharp edges on either the blades or the body of the wind turbine.	By looking at similar designs	Yes
13	The cost of the parts for the wind turbine shall be less than £20.00.	Adding up the cost of the parts in the material list	Yes
14	The total cost of the wind turbine including labour shall be less than £100.00.	Adding up the material cost from 13 and looking at the prices of similar products	Probably

Activity

Create a list of all the different forms of media that might be used during a presentation to the client. For each different form, list its advantages and disadvantages.

Summary

The final design proposal must normally be presented to the client before it proceeds to production.

A variety of different media will often be used to show the client how well the design meets the needs of the brief and the specification.

he client may request changes to the design proposal.

End of Chapter Review

Once a company has decided that the development of a new product is justified, they prepare a design brief. This is a short statement that tells the designers what is required. The designers will analyse the brief and carry out research into what is needed. From this, they will develop a specification, listing all the needs that the product must meet.

They will then generate some design ideas. They may use sketches to capture the ideas or develop them through the use of models. The needs will be compared with the specification to see how well the possible designs meet the requirements.

The ideas with the most potential will be further developed; drawings might be produced of what the finished design proposal might look like using rendered isometric projection. Computer Aided Design might be used to produce 3-D models to test how the components fit together. Other CAD packages might be used to develop and model the working parts of the system. Systems diagrams, circuit diagrams and flowcharts might be created to identify what is needed in the product and how it will work.

Orthographic drawings will be produced, either manually or using CAD, to communicate the dimensions of the design proposal to the manufacturing team. So that these can be understood by any engineer, it is important that they conform to all of the relevant standards and conventions. Assembly drawings and exploded views might be created to assist in the manufacture and maintenance of the product.

The final design proposal will be presented to the clients for approval. The clients will be shown how well the product meets the brief and the identified needs, along with the drawings of the proposed product and, for functional products, details about how it will work. The clients may request changes to the design proposal, if they think that this may allow the finished product to better meet the needs of the market. They will then decide whether to proceed to production.

∞ links

Further information on steps in the design process
www.techitoutuk.com/knowledge/designprocess.html and
www.bbc.co.uk/schools/gcsebitesize/design/electronics/elecdesignrev1.shtml

Further information on sketching techniques
www.bbc.co.uk/schools/gcsebitesize/design/graphics/drawingsketchingrev1.shtml

Further information on simple rendering techniques
www.technologystudent.com/despro2/drawtec2a.htm

Further information on isometric drawing
www.technologystudent.com/despro_flsh/isomty2.html

Further information on orthographic drawing
www.deyes.sefton.sch.uk/technology/Keystage4/orthographic_drawing.htm

Further information on CAD
www.cadinschools.org/index.php

Further information on systems diagrams
www.deyes.sefton.sch.uk/technology/Keystage4/systems_blocks.htm

Further information on electronic components and producing circuit diagrams
www.bbc.co.uk/schools/gcsebitesize/design/electronics/materialscomponentsrev1.shtml

AQA Examination-style questions

1 Write a design brief for a lawnmower. This should include a statement of the role to be carried out, the potential user, any important features and a manufacturing constraint. *(4 marks)*

2 Explain what is meant by the term 'design constraint'. Give an example of a constraint that would need to be considered during the development of a bicycle for use in the Olympics. *(2 marks)*

3 Describe two different ways in which modelling can be used during the design of a product. *(4 marks)*

4 A mobile phone manufacturer is thinking about launching a range of phones using a sports theme. Produce three labelled sketches showing different design ideas.
(3 marks for the sketches, 3 marks for labelling and 3 marks for the quality of the sketching)

5 Create an isometric drawing of a toolbox. All important features should be labelled.
(2 marks for correct isometric form, 2 marks for labelling, 2 marks for the quality of the drawing).

6 What is the meaning of the following lines in an orthographic drawing?

(a) ▪ ▪ ▪ ▪ ▪ ▪ ▪ ▪ ▪   (b) ────────   (c) ▪ ▪ — ▪ ▪ — ▪ ▪ — ▪ ▪
(3 marks)

7 What is meant by 'scale' on an orthographic drawing? *(1 mark)*

8 What is the meaning of the following symbol?

(1 mark)

9 List three conventions that must be used in an orthographic drawing. *(3 marks)*

10 Describe what is meant by CAD. Give two examples of how CAD is used during the design of a product. *(4 marks)*

11 List 4 advantages of producing drawings using CAD over producing drawings by hand. *(4 marks)*

Double award

12 Explain what is meant by a 'design specification' and describe how this could be used to evaluate a product. *(2 marks)*

13 A company that sells sports equipment wants to design a float for use in a carnival. Produce a rendered sketch of a possible design idea. Label any important features.
(2 marks for the rendered sketch, 2 marks for labelling and 2 marks for the quality of the sketching)

14 Explain the differences between an assembly diagram and an exploded view and how each of these is used. *(4 marks)*

15 A simple doorbell includes a battery, a switch, a buzzer and connecting wires. Draw the circuit diagram.
(2 marks for circuit, 3 marks for symbols and 1 mark for quality of drawing)

16 Create a systems diagram for an MP3 player. The parts of the systems should be clearly labelled. *(3 marks for the diagram, 4 marks for the labels and 1 mark for quality of drawing)*

17 Draw a flowchart to control the lights at a pedestrian crossing. It must allow the lights to change so that people can cross the road. Use standard flowchart symbols.
(8 marks for correct flow shown, 2 marks for symbols)

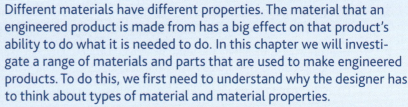

Objectives

List the main types of material.

Explain how a range of material properties relate to the design needs.

List manufacturing issues affecting the material selection.

Different materials have different properties. The material that an engineered product is made from has a big effect on that product's ability to do what it is needed to do. In this chapter we will investigate a range of materials and parts that are used to make engineered products. To do this, we first need to understand why the designer has to think about types of material and material properties.

■ Types of material

Materials are normally grouped together into types based on what they were made from. This gives five main categories:

- Metals
- Polymers
- Ceramics
- Composites
- Wood (which is not considered in this course)

Recently, some materials have been developed with properties that can change in response to changes in their environment. Although each of these materials falls into one of the first four types, they are sometimes referred to as a separate group called 'smart' materials.

We often think that each material type has certain characteristic properties. For example, we might believe that all metals are difficult to break, all thermoplastics bend easily or that all ceramics smash when dropped. However, while there may be some typical characteristics within any type, there will be big differences between the properties of individual materials. When choosing the best material for a part the material types can provide a starting point, but the designer must consider the properties of individual materials.

■ Material properties

The designers have to select materials with properties that meet the needs of the product. To do this, they look at the product specification. For each need, they will identify the properties that the material used to make the product must have. Examples of the questions that the designers may ask and some of the material properties that they might consider are shown in Table **A**.

Many products need the material to have a combination of different properties. It can be difficult to identify a material with exactly the right combination, so the designers often have to compromise. For example, they may choose a less attractive material because it is stronger. Alternatively, they may change the design so that it uses a weaker material if it is cheaper.

A *Considering what material properties the design needs*

Design question	Material property
Does the product have to withstand forces, such as loads being put on it, being pulled or being twisted?	Strength
Does the product need to be resistant to scratches and wear?	Hardness
Does the material need to be resistant to knocks and bumps?	Toughness
Does the product have to be light, so that it can be moved easily or carried around?	Weight
Does the product need to be in a certain price range?	Cost
Does the product have to work in an environment that could damage it?	Corrosion resistance
Should the product be a particular colour, texture or style?	Aesthetics
Does the product need to be recycled?	Sustainability
Does the material need to allow electricity to pass through it?	Resistivity
Does the material need to stop heat from passing through it?	Thermal conductivity

■ Material considerations for making the product

As well as considering the properties of the materials that are needed for the product to function as required, the designers must also think about how the product is going to be made. Their choice of materials may be limited for a number of reasons. For example:

- The 'ideal' material may not be available in the amounts needed locally or may need to be ordered from specialist suppliers.

- The equipment that is available to manufacture the parts may only be suitable for some types of material. This will be covered in more detail in Chapter 4.

- To simplify costs and manufacture, the designers may also consider using 'standard' parts for some components in the product, rather than making all of the parts to the exact sizes needed.

- Alternatively, they may have to add extra materials properties to the list of needs. For example, they may have to consider the malleability of the material, which means how easy it is to shape, spread or form.

B *Gears can be made from a variety of different materials*

Starter activity

Create a list of the material properties that would be required for the following applications:

- The engine for a racing car
- The case of a games console

Metal is made from metal ore. This has to be mined, refined and processed to turn it into a usable form. Metals are commonly available in a wide range of shapes and sizes, Diagram **A**. This reduces the amount of work that might be needed to change their shape or form.

It is rare for a metal to be used in its pure form. Normally they are mixed with other metals to improve their properties. A mixture of two or more metals is called an **alloy**. The proportion of the other metals added to form an alloy can typically range from 0.1% up to 50%.

There are two main types of metal: **ferrous** and **non-ferrous**. Ferrous metals contain iron. Non-ferrous metals do not contain iron. Both types of metals can be recycled.

Objectives

Explain what is meant by alloy, ferrous and non-ferrous metal.

Describe the properties and typical uses of ferrous metals.

Describe the properties and typical uses of a range of common non-ferrous metals.

▮ Ferrous metals

Pure iron is generally too soft for most engineering products. Adding carbon makes a new material called carbon steel, which is probably the most widely used metal in the world. The amount of carbon added has a significant effect on the material properties, Table **B**.

Most carbon steels have poor corrosion resistance, which means that they rust easily. They are often coated with other metals, such as zinc, or painted to improve their corrosion resistance.

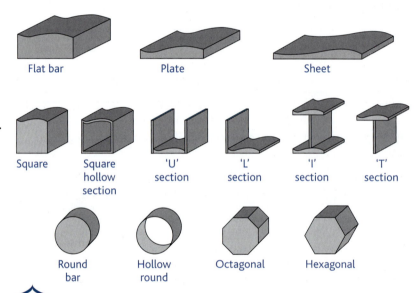

Flat bar Plate Sheet

Square Square hollow section 'U' section 'L' section 'I' section 'T' section

Round bar Hollow round Octagonal Hexagonal

A *Examples of forms that metals are available in*

B *Common uses of some metal casting*

Metal	Alloying elements include:	Properties	Examples of what it is used for
Low carbon steel	Up to 0.3% carbon	The weakest of the steels but still stronger than most non-ferrous metals. Easy to machine, tough and cheap. Cannot be hardened or tempered.	Car bodies, panels for fridge doors, steel sheet, screws, nails
Medium carbon steel	0.3–0.7% carbon	Strong, tough, hard and fairly cheap; more difficult to form than mild steel. Cannot be hardened.	Gears, railway wheels, high strength tubing
High carbon steel	0.7–1.4% carbon	Stronger and harder than medium carbon steel, but not as tough. Difficult to form but quite cheap. Can be hardened and tempered.	Ball bearings, knives, saws, chisels, hammers
Cast iron	Typically 3–3.5% carbon	Good compressive strength. Hard but can be brittle. Melts easily, so very good for casting. Cheap. Poor corrosion resistance – rusts easily.	Engine blocks, engineering vices, anvils
Stainless steel	At least 11.5% chromium	Hard and can be difficult to machine. Quite expensive. Good corrosion resistance – won't rust.	Sinks, kettles, medical equipment, knives and forks

Common non-ferrous metals

There are many useful metals and alloys that do not contain iron. Most of the common non-ferrous metals have good corrosion resistance and do not require surface finishes. Common non-ferrous metals include:

- aluminium
- titanium
- zinc
- pure copper
- bronze and brass.

Aluminium is one of the most commonly occurring materials in the Earth's crust. It has excellent corrosion resistance and a lower density than steel, so finished products of the same size are lighter in weight. However, aluminium is more expensive than steel and it is not as strong, so it is often alloyed to improve its uses. Uses for aluminium and its alloys include aircraft structures and cans for soft drinks.

Titanium is stronger that aluminium, but it is more expensive. Like aluminium, it has excellent corrosion resistance and a lower density than steel, so finished products of the same size are lighter in weight. Titanium and its alloys are used for high performance aircraft structures.

Zinc has a low melting point, which makes it ideal for die-casting. It can be alloyed with aluminium to increase strength. Common uses for zinc are camera bodies, handles for car doors and components for cars, such as carburettors and fuel pumps.

Pure copper is a very good conductor of electricity and is flexible but it lacks strength. It is commonly used to make wires and pipe-work for domestic use. Copper oxide is sometimes added to copper to make it stronger.

Bronze is an alloy of copper and tin and is frequently used for casting. Brass is an alloy of copper and zinc, which is difficult to cast but can be machined to a good finish. The uses of bronze and brass include high pressure valve bodies, statues, doorknobs and electrical parts.

C   The stainless steel body of a toaster

D   A bronze statue

Key terms

Alloy: a mixture of two or more metals.

Ferrous metal: a metal that contains iron.

Non-ferrous metal: a metal that does not contain iron.

Activities

1   Research as many different types of non-ferrous alloy as you can. Make notes of their names, composition and typical uses.

2   Recommend a suitable metal to make each of the following parts of a mountain bike. Provide an explanation for your choice, by comparing it to other metals: the pedals, the frame, the wheel rim and spokes.

AQA Examiner's tip

When discussing the selection of materials, make sure that you describe the most important properties for the context, e.g. weight and strength for aircraft construction.

Summary

An alloy is a mixture of two or more metals.

Carbon steels are relatively cheap and have good strength, but poor corrosion resistance. They are used in a wide variety of applications.

Common non-ferrous metals include aluminium, copper, zinc, bronze and brass. While these are not typically as strong as steel, they have other properties that make them suitable for many applications.

Materials and their properties: polymers

What are polymers?

Polymers are the most widely used materials in commercial production. The name 'polymer' comes from the words 'poly', meaning many, and 'mer', meaning parts. They are manufactured by a process called polymerisation. This involves joining monomers together to form long chains of molecules. So, for example, polypropylene is made up of single monomers of propylene, joined together to form a long chain. A polymer product may contain a large number of the long chains.

Polymers can be created from two main sources:

- synthetic polymers
- natural polymers.

Synthetic polymers are by far the most commonly used. These are chemically manufactured from carbon-based materials, such as crude oil.

Natural polymers are made by processing natural materials, such as plants. The most widely used natural polymer is latex from trees, which is used to make a form of rubber. As environmental concerns increase and oil reserves reduce, there is increasing interest in developing new polymers made from natural sources.

There are two main types of polymer: thermosetting polymers, also called thermosets, and thermoplastic polymers, also called thermoplastics. Within these types, there are many different varieties of polymer with very different properties. Polymers vary considerably in strength and some of the stronger compare favourably with metals. All polymers have a lower density than metals, which means that a product of the same size will weigh less than one made out of metal. Polymers are not normally painted, but their colour can be changed by adding pigments to them.

Thermosetting polymers

Thermosets are typically formed by a moulding process. During the process, they form many links across the different polymer chains, which stop the chains being able to move. This makes them harder and more rigid than thermoplastics, with good resistance to electricity and heat. Once moulded, they cannot normally be re-shaped and they cannot be recycled. Commonly occurring thermosets include:

- Melamine formaldehyde. Its common uses include impact-resistant plastic plates, laminate coverings for kitchen workshops and cupboards, and pale-coloured mouldings for electrical equipment, such as lighting fittings.
- Phenol formaldehyde. A dark coloured, brittle plastic used to make electrical fittings such as light sockets, and heat-resistant parts for domestic appliances, such as handles for cookers and pans.
- Urea formaldehyde. This can be coloured with artificial pigments and is used to make electrical fittings, such as light switches.
- Epoxy resin. This is used to make printed circuit boards and can be 'cold cast' to form electrical insulators.

Objectives

Explain what is meant by polymer, thermoset and thermoplastic.

Describe the properties and typical uses of thermosetting polymers.

Describe the properties and typical uses of a range of thermoplastics.

Key terms

Polymer: an organic material made up of a chain of single units called monomers.

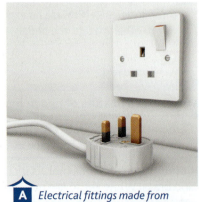

A *Electrical fittings made from thermosets*

Thermosets are normally available in powder or granular forms. They may be mixed with fillers and colouring agents. A filler is a cheap material used to provide extra bulk and reduce the amount of polymer needed. This reduces the cost of the thermoset and modifies the properties of the material.

Thermoplastic polymers

Thermoplastics do not have links between the different polymer chains. This means that they are softer and more flexible than thermosets. They soften when heated and can be shaped when hot. The shape will harden when it is cooled, but can be reshaped when heated up again.

Thermoplastics can normally be recycled. They are readily available in sheets of standard thicknesses or in granular forms for use with moulding processes. There are a wide range of thermoplastics in common use – see Table **C**.

B *Drinks bottle made from thermoplastic*

C *Typical uses of some common thermoplastics*

Metal	Properties	Examples of what it is used for
Polypropylene	Softens at 150°C. High impact strength for a polymer. Can be flexed many times without breaking.	Food containers, plastic chairs, children's toys
High impact polystyrene (HIPS)	Softens at 95°C. Easy to mould. Light but strong.	Vacuum formed packaging and casings
Acrylic (polymethyl methacrylate – PMMA)	Good optical properties. It can be transparent, like glass, or coloured with pigments. Hard wearing and will not shatter on impact.	Machine guards, plastic windows, bath tubs, display signs
Nylon	Low friction qualities. Good resistance to wear and tear.	Bearings, gear wheels, curtain rail fittings
High density polyethylene (HDPE)	Softens at 120°C. Strong.	Bowls, buckets, milk crates
Low density polyethylene (LDPE)	Softens at 85°C. Softer, more flexible and less strong than HDPE.	Detergent bottles, carrier bags, packaging, film
Polyvinyl chloride	Stiff and hard wearing. Can be made softer and rubbery by adding a plasticiser.	Chemical tanks, pipework, coverings for electric cables, floor and wall coverings, packaging

Activities

1   Some plastics are marked with a symbol containing a number, to show that they can be recycled. These were explained in Topic 1.9. Identify an example of a product made with each of the different types of polymer that can be recycled.

2   Recommend a suitable polymer to make the cover for an MP3 player. Provide an explanation for your choice, by comparing it to other polymers.

Summary

Thermosetting polymers cannot be reshaped or recycled. They are often used in applications where a combination of strength and resistance to electricity or heat is needed.

Common thermoplastics include polypropylene, nylon and acrylic. These can be softened and reshaped when heated.

AQA *Examiner's tip*

You should be able to describe the unique properties of the two types of polymer and compare them to other materials using several parameters, e.g. strength, weight, cost.

◼ Ceramics

When we think of ceramics, often the first things that come to mind are glasses, cups, saucers and pots. The name **ceramic** even comes from a Greek word meaning 'potter's clay'. However, ceramic materials are also used for a wide range of different engineering applications.

Examples of uses of ceramic materials

- tools for grinding and cutting
- refractory tiles used to insulate furnaces
- electrical insulators, such as the alumina casings on spark plugs
- lenses
- building materials, such as plaster, cement and bricks.

Ceramic materials are often oxides, nitrides or carbides of metals. They have excellent corrosion resistance, generally have good strength in compression and are harder than most other engineering materials. This means that they are very resistant to scratches and wear. However, they tend to be weak in tension and brittle – this means that if they are pulled they don't stretch, but tend to crack and fall apart.

Ceramics are good insulators against both heat and electricity. They can normally withstand high temperatures without softening, so they are used in preference to polymers in applications where electrical insulators are needed and where flexibility is not required.

Ceramic materials are difficult to machine due to their hardness and the risk of breaking. This means that ceramic components are frequently made by moulding processes, often using high temperatures, so that machining is not required. Many ceramic materials could be recycled in principle. However, with the exception of glass, it is not normally cost effective to recycle ceramic materials.

Most engineering ceramics are used in polycrystalline form. This means that they have a structure made up of a lot of very tiny crystals of the ceramic material. There is a special group of ceramics called semi-conductors. These are grown as very pure single crystals under carefully controlled conditions. The full size crystals range up to 150 mm in diameter and 2500 mm in length. These are cut into thin wafers and made into devices such as integrated circuits, transistors and diodes. Semi-conductors are used in every modern electronic devices, such as computers, mobile phones etc.

> ### Objectives
>
> Describe the properties and typical uses of engineering ceramics.
>
> Describe the properties and typical uses of a range of composites.

> ### Key terms
>
> **Ceramic:** an inorganic material, normally an oxide, nitride or carbide of a metal.
>
> **Composite:** a material that is made from two or more material types that are not chemically joined.

A *A ceramic furnace lining*

Composites

A **composite** is a material that is made by combining two or more different types of material. The materials are not joined chemically – they still remain physically distinct. That means that, for example, in fibreglass, if you looked at its structure under a microscope you would be able to see fibres of ceramic material, surrounded by polymer. Composites combine the properties of the materials that they are made from.

The benefit of composites is that you can create materials with particular, unique combinations of properties. This means that you can make products which were not previously possible, such as lightweight bulletproof armour. However, there is a major disadvantage in that most composites cannot be recycled as it is very difficult to separate the two materials. Some of the most commonly occurring composites are:

- reinforced concrete
- fibreglass
- carbon-reinforced plastic
- metal matrix composites.

B *Canoes made from composite materials*

The most commonly used composite material is reinforced concrete. Like many ceramic materials, concrete has good compressive strength but poor strength under tension. Steel has very good strength under tension. By using steel rods to reinforce concrete, the finished material has the compressive strength of concrete and greatly increased strength under tension.

Fibreglass uses thin fibres of ceramic glass stuck together with a polyester or epoxy resin. It is a very strong material that can be formed into complex shapes. It is normally made by positioning sheets of fibre material in a mould and soaking it in the resin. After it has dried and hardened, the finished product is often painted.

Carbon-reinforced plastic (CRP) products are stronger than fibreglass. They are manufactured in the same way and used to make high performance items such as racing car bodies, helmets and bulletproof vests.

Metal matrix composites are made by dispersing a reinforcement material into a metal matrix. For example, carbon or ceramic fibres are commonly used in an aluminum matrix to make a composite which is lightweight and has high strength. This is used for drive shafts and cylinder liners in high performance cars.

AQA Examiner's tip

You should be able to describe the unique properties of ceramics and composites and compare them with other materials using several parameters, e.g. strength, weight, cost.

Summary

Ceramic materials have excellent corrosion resistance, and are good insulators to both heat and electricity. Although they are amongst the hardest engineering materials, they tend to be weak in tension and are brittle.

Composites are a combination of two or more different types of material. They combine the properties of the materials that they are made from.

Activities

1. Choose a product made from a ceramic material and carry out a product analysis.

2. Research other uses of ceramics, not discussed in this topic. Explain reasons for their use and the properties required.

3. Recommend a suitable material to make the frame for a high performance sports bicycle. Provide an explanation for your choice, by comparing it to other materials that you might have chosen.

3.4 Smart materials

Smart materials have properties that can change in response to changes in their environment. This means that they have one or more properties that can be changed by an external condition, such as temperature, light, stress or electricity. This presents the designer with some exciting options to consider for new products in the future. Some of the smart materials that are already finding uses include:

- shape memory alloys
- viscoelastic compounds
- piezoelectric materials
- quantum tunnelling composites
- colour change materials.

Shape memory alloys

Most materials show some, but limited memory due to elasticity. When stretched a little they can spring back to their old shape. However, when stretched further or bent, they stay that way. If a part made from a **shape memory alloy (SMA)** is bent out of shape, when it is heated above what is known as its transition temperature, it will return to its original shape. This cycle of bending and being straightened can be repeated many times. SMAs can be formed into almost any shape, from springs to flat plates, and be conditioned to return to this shape by heating them above their transition temperature.

The most common SMA is an alloy of the metals nickel and titanium, which has a transition temperature of 70°C. The heating can be achieved through direct means or by passing an electric current through it.

Viscoelastic compounds

Viscoelastic compounds are a type of smart fluid. They are normally soft and mouldable. However, if they are suddenly impacted they act like rubber and bounce. In addition to being used to make toys like Silly Putty, this type of material can be used to change the response of vehicle suspension systems.

Piezoelectric materials

Piezoelectric materials do not conduct electric current. However, when squeezed rapidly they produce an electrical voltage for a moment. Alternatively, if a voltage is put across the material there is a very small change in shape.

An example of a natural piezoelectric material is quartz. A wide range of materials with piezoelectric behaviour have been developed, including polymers and thin-film ceramics. These have found many uses, including:

- contact sensors for alarm systems
- microphones and loudspeakers

Information

Applications of shape memory alloys

- Triggers to start the sprinklers in fire alarm systems
- Controllers for hot water valves in showers or coffee machines
- Shrink-fit seals for hydraulic tubing
- Artificial muscles in robot hands
- Spectacle frames

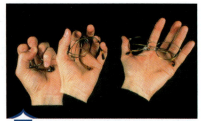

A *Spectacle frames made from shape memory alloy*

- electrical generators
- motors to move the lenses on cameras
- regulators for electric circuits, such as quartz clocks.

Quantum tunnelling composite

Quantum tunnelling composite (QTC) is a flexible polymer which contains tiny metal particles. In its normal state it is an insulator, but when squeezed it becomes a conductor able to pass high currents. QTC can be used to make membrane switches like those used in mobile phones, pressure sensors and speed controllers.

Colour change materials

Thermochromic materials change colour as the temperature changes. The colour changes are based on liquid crystal technology. At specific temperatures, the liquid crystals, which may be as small as 10 microns in diameter, reorient their structure to produce an apparent change of colour.

Examples of the use of these materials include:

- plastic strips that use colour changes to indicate temperature or act as thermometers
- test strips on the side of batteries. These heat a resistor printed under the thermochromic film. The heat from the resistor causes the film to change colour.
- packaging materials that show you when the product they contain is cooked to the right temperature.
- colour indicators on cups, to show whether the contents are hot.

Photochromic materials change colour according to different lighting conditions. They are particularly reactive to ultraviolet light. They are used for products ranging from nail varnish, security markers that can only be seen in ultraviolet light, jewellery and mobile phone cases.

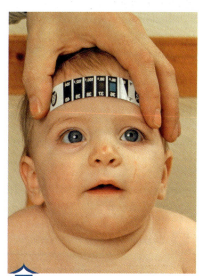

B *Contact thermometer made from thermochromic sheet*

Activities

1 Smart materials are being developed which can change properties when subjected to a magnetic field. Identify what these materials are made from and their potential applications.

2 Choose one type of smart material. Write a short report explaining which other materials it could be used to replace, explaining its potential advantages and drawbacks.

Summary

Smart materials can change one or more of their properties in response to changes in their environment.

There are smart materials which can change shape with temperature, change electrical resistance with stress or change colour with temperature or light levels.

AQA Examiner's tip

Use real life examples when describing the use of new and smart materials. Say why they have an advantage over traditional materials.

3.5 Standard parts and components: electrical and electronic

As well as thinking about the properties of the material, the designer has to take account of manufacturing and cost limitations. One key decision is whether or not to use standard parts. These are **components** that are made to a known specification or tolerance, which are bought from suppliers. The suppliers have the facilities to make large numbers of these parts at low cost and consistent quality. The disadvantage of standard parts is that they are only available in set values. This means that the designer may have to adjust the design to accommodate them or use a combination of standard parts to achieve a required result. However, it is still normally much more cost effective to use standard parts than to make customised parts.

When designing using standard parts, symbols can be used rather than drawing the part. Some of the most commonly used standard parts include:

- electrical and electronic components
- pneumatic and hydraulic components
- mechanical parts and fasteners.

Electrical and electronic components

Electrical components can be grouped according to how they are used in a circuit. It is easiest to understand these functions if we think of the circuit as being a system (see Diagram **A**). The main types of function are inputs, processes, outputs and supporting parts.

Inputs

Input components activate the circuit. They include switches and sensors, Table **B**.

A Systems diagram

B	*Examples of some common input components*	
Component	**Symbol**	**Use**
Switch (latching)		Can be used to turn the circuit on. Stays on until turned off.
Switch (push to make)		Can be used to turn the circuit on or as a contact sensor. Only stays on while pressed.
Thermistor		Used for temperature sensing. It is a type of resistor where the resistance increases as the temperature increases.
Light dependent resistor (LDR)		Used for sensing light. It is a type of resistor where the resistance increases as the amount of light increases.

C	*Examples of some common process components*	
Component	**Symbol**	**Use**
Integrated circuit		There are lots of different types of integrated circuits, ranging from timers to logic gates to programmable chips. Used for amplifiers, counters, timers, processors and oscillators.
Transistor		An electronic switch. Can be used as part of an amplifier to boost the power of an output device.

D *Examples of some common output components*

Component	Symbol	Use
Light Emitting Diode (LED)		Converts electricity to light and only allows the electricity to pass through one way. Uses much less electricity than a bulb.
Lamp	⊗	Turns electricity into light
Buzzer		Turns electricity into sound. Makes simple buzzing noises only
Speaker		Turns electricity into sound
Motor	Ⓜ	Turns electricity into circular movement

E *Examples of some common supporting components*

Component	Symbol	Use
Battery		A collection of cells, used to provide power to a circuit
Resistor		Can be used to control the amount of current in a circuit or to protect other components
Variable resistor		A resistor whose resistance can be changed. Can be used to control the current in a circuit
Fuse		A special type of resistor that will burn out if the current goes above its rated level. Used to protect circuits from too much current
Capacitor		Used to store charge. Can be used in conjunction with a resistor to provide a simple timing circuit
Diode		Only allows electricity to flow in one direction. Can be used to protect components
Relay		Can be used as part of a low-powered circuit to turn on a higher-powered circuit.
Wire	———	Used to join other components

Processes

Process components control what the circuit does in response to the input, Table **C**. For example, it could be that the process component turns on the outputs only if two sensors are activated or only turns on the outputs for a set period of time. Many of these devices are **electronic components**.

Outputs

Output components include a range of devices that turn electricity into light, sound or movement, Table **D**.

Supporting parts

There are a number of components that can be used in any of the above systems functions, to support or customise the working of the systems functions, Table **E**.

Activities

1   Research and find at least three electrical circuit diagrams that use some of the above components. Explain the function of each of the components in the circuit.
2   Locate pictures of all of the electrical components listed in the above tables.

Summary

Standard parts are used because it is cheaper to buy them than to make them in small quantities. If they go wrong they can be easily replaced.

A wide range of standard electrical and electronic components are available. Each carries out a specific function.

Key terms

Component: a part used to make up an assembly.

Electrical components: devices that are simple conductors that allow current to flow through them.

Electronic components: devices including semi-conductor materials in an electrical circuit.

AQA Examiner's tip

You should be able to identify a range of electrical components and the symbols used to describe them.

3.6 Standard parts and components: pneumatic, hydraulic and mechanical

Pneumatic and hydraulic components

Pneumatic systems are powered by gas, normally air, instead of electricity to carry out their function. Similarly, **hydraulic systems** use liquid instead of electricity. Both types of system rely on moving pressure from one place to another. The gases used in pneumatic systems can be compressed and the pressure this creates is used to do work. Liquids do not compress and are used in systems where much heavier work is required.

The reasons why these systems may be used in some applications instead of electric circuits include:

■ It might not be safe to use electricity due to the risk of sparks or possible interactions with other electrical equipment. For example, a common use of pneumatic systems is to clamp tools, open doors or move parts in electrically-powered machine tools.

■ It might be easier to carry out the task needed by the direct force rather than converting electricity to movement. For example, a common use of hydraulic systems is to push the brake pads in a car onto the brake discs.

Similar to electrical and electronic components, pneumatic and hydraulic components are represented by symbols, Table **B**. On a drawing, the difference between a pneumatic or hydraulic circuit is that pneumatic systems use arrows that have open heads and hydraulic systems use arrows that have solid heads, Diagram **A**.

Pneumatic arrow   Hydraulic arrow

A *Arrow types for hydraulic and pneumatic systems*

B *Examples of some common pneumatic and hydraulic components*

Component	Symbol	Use
Push button operation		Can be pushed to turn the circuit on. Only stays on while pressed
Lever operation		Can be used to turn the circuit on. Stays on until turned off
3 port valve		Can be used to direct the gas or liquid from an input port in one of two different directions depending upon whether the push button on it has been pressed. Used to activate different parts of the circuit
5 port valve		In effect, a 'double' version of the 3 port valve, but still controlled by a single button
Shuttle valve		Can be used as an 'or' gate, to activate the circuit in response to either of two connected inputs
Single acting cylinder with spring return		Turns the air pressure into linear force or movement when pressure is applied. The spring return automatically closes the cylinder when the pressure is removed.
Double acting cylinder		Turns the air pressure into linear force or movement, which can be used to both open or close the cylinder.

Component	Symbol	Use
Reservoir		Used to hold the gas (pneumatic system) or liquid (hydraulic) system
Flow regulator/ unidirectional flow control valve		Used to control the amount of gas or liquid flowing and allows the gas or liquid to flow in one direction only
Pipes		Allow the gas or liquid to flow round the system
Mains air supply		Shows the location of the pump used to power a pneumatic system
Exhaust air		Allows the gas in a pneumatic system to escape, removing the pressure

◼ Mechanical parts

Mechanical parts are components that move in machines or equipment, or are the parts that can be used to make moving equipment. A good test of whether a mechanical part is a standard part is to ask if it could be used without modification during the manufacture of a different piece of equipment. If it cannot be reused, it is probably a customised part. There are a wide range of different standard parts that a designer can opt to use, Table **D**.

Example of a product using pneumatics

D *Examples of some common mechanical parts*

Standard parts	Use
Threaded fixers	Screws, nuts and bolts (see Topic 4.7)
Non-threaded fixers	Split pins, circlips and wire rings – hold components in place
Springs	A coil of metal – often used to allow movement within a device, such as opening and closing a valve
Bearings	Can be metal or plastic – used to hold a rotating part in place but allows it to move. Bearings are an important component of most moving machinery, such as lathes and cars.
Gears	Can be made from metal or plastic – used to change the speed or torque of a motor
Belts and pulleys	Used to join moving parts

Standard parts are generally listed in parts catalogues. They will most likely have a range of set sizes. It normally costs much less to change the design to accommodate standard parts than to make customised parts. A second benefit of standard parts is that if they break or wear out, they can be easily replaced. This means that the equipment can be easily repaired, rather than having to replace the product or wait while a customised part is made. This also means that standard parts can be more environmentally friendly than using customised parts.

AQA/ Examiner's tip

You should be able to identify a range of pneumatic components and the symbols used to describe them.

When selecting a component state why it is suitable and if another component could have been used instead.

Summary

A range of standard pneumatic and hydraulic components are available. Each carries out a specific function.

A wide range of standard mechanical parts are available, including fixings and moving parts such as bearings and gears. It is typically cheaper to buy these components than to make them in small quantities.

Activities

1 Research and find at least three pneumatic or hydraulic circuit diagrams that use some of the above components. Explain the function of each of the components in the circuit.

2 Carry out an analysis of an engineered product with moving parts and identify all of the standard parts.

Selecting the materials to make the product

Chooser charts

Table **A** compares some of the properties of a range of different materials. This is a subjective comparison and the ratings will vary for different types of application. A designer might use a chooser chart like this to identify the materials with the combination of properties that may satisfy the needs of an application. In creating this list, he will also have to consider aesthetic issues, such as what the material looks like, and manufacturing issues such as materials availability and the ability of the available equipment to process each material. Once he has identified possible materials, he will consider the quantitative properties against the design needs – that is, how the results of materials testing compares against the calculated needs of the product.

Objectives

List the comparative properties of different materials.

Explain how a chooser chart can be used to assist the selection of materials for an application.

AQA Examiner's tip

You should be able to compare the properties of different materials using several parameters, e.g. strength, weight, cost.

A Subjective comparison of material properties

Material	Comparative properties								
	Strength	Hardness	Toughness	Weight	Corrosion resistance	Ability to conduct electricity	Ability to conduct heat	Ability to be recycled	Cost
Low carbon steel	Very good	Good	Very good	High	Poor	Very good	Very good	Excellent	Low
Medium carbon steel	Excellent	Very good	Excellent	High	Poor	Very good	Very good	Excellent	Low
High carbon steel	Excellent	Very good	Very good	High	Poor	Very good	Very good	Excellent	Medium
Cast iron	Good	Good	Low	High	Poor	Very good	Very good	Excellent	Low
Stainless steel	Excellent	Very good	Very good	High	Good	Very good	Very good	Excellent	High
Aluminium alloys	Very good	Good	Very good	Low	Very good	Very good	Very good	Excellent	Medium
Titanium alloys	Excellent	Very good	Excellent	Low	Very good	Very good	Very good	Excellent	High
Zinc	Good	Good	Very good	Medium	Good	Very good	Very good	Excellent	Medium
Copper	Medium	Medium	Good	High	Ok	Excellent	Excellent	Excellent	High
Brass	Good	Good	Very good	High	Good	Very good	Very good	Excellent	High
Melamine formaldehyde	Medium	Very good	Low	Medium	Very good	Very poor	Very poor	Very poor	Medium
Phenol formaldehyde	Medium	Good	Low	Medium	Very good	Very poor	Very poor	Very poor	Medium
Polypropylene	Low/ medium	Low	Good	Low	Very good	Very poor	Very poor	Excellent	Low
HIPS	Medium	Low	Good	Low	Very good	Very poor	Very poor	Very good	Low
Acrylic (PMMA)	Medium	Low	Good	Low	Good	Very poor	Very poor	Very good	Medium
Nylon	Good	Good	Good	Low	Very good	Very poor	Very poor	Very good	Low
HDPE	Low	Low	Good	Low	Very good	Very poor	Very poor	Excellent	Low
LDPE	Low	Low	Good	Low	Very good	Very poor	Very poor	Excellent	Low
PVC	Low	Good	Medium	Low	Very good	Poor	Very poor	Good	Low
Alumina	Very good	Excellent	Poor	Medium	Excellent	Very poor	Very poor	Low	High

Alumina silicates	Good	Excellent	Poor	Medium	Excellent	Very poor	Very poor	Low	Medium
Fibreglass	Very good	Good	Very good	Low	Very good	Very poor	Poor	Very poor	High
CRP	Very good	Medium	Excellent	Low	Very good	Poor	Poor	Very poor	
Metal matrix composites	Excellent	Good	Excellent	Medium	Very good	Good	Good	Poor	Very high
Shape memory alloy	Very good	Good	Very good	Medium	Good	Very good	Very good	Very good	High

Materials selection: helicopter frame

Case study

The frame of a helicopter is the skeleton that the outside panels and engine are attached to. Some of the key material properties needed for this application are:

■ Its strength must be at least very good. It needs to support both its own weight and the high forces that have to be generated by its rotor blades.

■ It must be as low weight as possible. The lighter it is, the faster the helicopter will be able to go and the more fuel efficient it will be.

■ It should have at least very good toughness. This is important so that it does not break if it has a heavy landing or if it hits a bird at high speed.

■ It should have good corrosion resistance, as it may have to be out in the rain.

Examining Table **A**, a shortlist of suitable materials would include aluminium alloys, titanium alloys, fibreglass and carbon-reinforced plastic (CRP). The titanium alloys would offer better strength and toughness than the aluminium alloys, but they would cost more. Similarly, CRP would offer better properties than the fibreglass.

In practice, all four of these materials have been used in this type of application. The deciding factor as to which is the most suitable for any particular manufacturer to use has often come down to the manufacturing issues. These include how easy the material is to join together, both as a structure to the body panels, and what manufacturing facilities are available at the company.

These types of manufacturing issue will be considered in detail in Chapter 4.

B *A helicopter*

Activities

Identify the material properties that are needed and recommend a suitable material for the bonnet of a high performance sports car. You will need to provide an explanation for your choices, by comparing your recommendations with other materials.

Summary

Different materials have a wide range of properties and different combinations of properties.

A chooser chart is a useful tool to help select materials that could be used to make a product.

3

When selecting the materials for a product, the designer has to identify all of the material properties that may be required. He also has to take account of the manufacturing requirements and capabilities.

There is a wide range of materials that may be considered. These include ferrous and non-ferrous metal alloys, thermosetting and thermoplastic polymers, ceramics and composites. Each of these types of material offer unique combinations of properties.

In addition to selecting the materials, the designer also has to consider whether to use standard parts in the product. Some of the most commonly used standard parts include electrical and electronic components, pneumatic and hydraulic components and mechanical parts and fasteners. The potential benefits of using standard parts are that it is cheaper to buy them than to make them in small quantities and if they go wrong, they can be easily replaced.

∞ links

Further information on materials
www.technologystudent.com/joints/joindex.htm and
www.bbc.co.uk/schools/gcsebitesize/design/resistantmaterials/materialsmaterialsrev1.shtml

Further information on smart materials
www.tep.org.uk/Frames/_f_millennium_smart.html

Further information on electrical and electronic materials
www.bbc.co.uk/schools/gcsebitesize/design/electronics/

Further information on pneumatics
www.bbc.co.uk/schools/gcsebitesize/design/systemscontrol/pneumaticsrev1.shtml

AQA↗ Examination-style questions

1   When selecting the materials for a new product, the designer considers the properties that will be needed by the product. However, his choice of materials may be limited by practical considerations. List two of the possible reasons that may limit the choice of materials in practice.   (*2 marks*)

2   The bodies of domestic fridges and washing machines are normally covered in panels made from mild steel. This has normally been coated with nickel and painted white. Give two reasons why this material is suitable.   (*2 marks*)

3   Recommend a material that could be used to make the case of a mobile phone. Give two reasons for your choice.   (*3 marks*)

4   Which material would you use to make a holder for a light bulb? Give an explanation for your choice.   (*2 marks*)

5   The structure and skin of a racing car could be made from an aluminium alloy, a titanium alloy, or a composite material such as CRP. List one potential benefit for using each of these materials.   (*3 marks*)

6   Draw the symbols for the following components:
    (a) switch (latching)   (b) light emitting diode   (c) buzzer   (d) battery   (*4 marks*)

7   Identify the following components from their photos:

(a)    (b)    (c)    (d)

(*4 marks*)

Double award

8   Gears used in train engines are often manufactured from medium carbon steel.
    Give two reasons why this material is used rather than cast iron.   (*2 marks*)
    Give two reasons why this material is used rather than a non-ferrous alloy.   (*2 marks*)

9   Which material would you suggest to use for the handle on a saucepan made from stainless steel? Give two reasons for your choice.   (*3 marks*)

10   Lunchboxes and food containers are often made from polymers. Explain two benefits of using polypropylene for this application, rather than a thermoset.   (*2 marks*)

11   Compared with using a non-ferrous metal alloy, what is a disadvantage of using a composite material such as CRP for the manufacture of the structure of an aircraft?

12   Give one advantage and one disadvantage of using shape memory alloys for the manufacture of spectacles, rather than brass or stainless steel.   (*2 marks*)

13   List two smart materials that can change colour, in response to different changes in their environment. For each material, list an application where it could be used.   (*4 marks*)

14   A simple alarm circuit contains the following components: a battery, a contact switch, a light emitting diode, a buzzer, connecting wires
    Draw a circuit diagram showing the alarm circuit.
    (*2 marks for circuit, 4 for symbols, 1 for quality of drawing*)

Objectives

Describe the relationship between the design, material selection and how a product is manufactured.

Explain how the manufacture of parts in small quantities can affect process selection.

Selecting the most appropriate equipment to make a product has a big effect on that product's performance and cost. In this chapter we will investigate the processes and tools that are used to manufacture parts. Process selection is influenced by the number of parts to be made, so this chapter will focus on making one-off parts or small quantities. Chapter 5 will extend this to look at the equipment used for making parts in large quantities. To do this, we first need to understand what the designer has to consider when selecting processes. We then need to understand how manufacturing one-off products and small quantities of parts can affect process selection.

Process selection and the designer

When developing a product, the designer has to think about the design, the properties of the materials and how the product is going to be made. These three are closely related. For example, the design needs will determine the properties needed from the materials and therefore the possible materials that can be used. The choice of material will limit the processes that can be used, as some materials can only be processed in certain ways. The process may affect the shape and dimensions that can be achieved, needing changes to the design and so on.

Whilst considering these, the designer also has to consider some important practical limitations. For example, it is often necessary that the design is made using the existing processes and equipment at the company, as new processes and equipment can be very expensive and require specialist skills to operate. Further, the selected material for a part may only be available in certain sizes, shapes and conditions. This will determine what needs to be done to it to achieve the required size and shape of the final product. Considering these limitations, the designer has to identify the processes to be used and the tools and equipment to carry out these processes.

Information

Examples of parts made as one-offs

- Designer jewellery
- Tailored clothes
- Cranes for large construction projects
- Aircraft carriers
- Replacements for broken parts in a machine
- A prototype of a new design of a car

Production in small quantities

One-off manufacture is the production of a single item. This could be a new item designed to a customer's individual specification or a replacement item made to an existing design. Sometimes one-off products are made to test how well a product design will work, before making the product in large quantities – these are called prototypes.

Small batch production is the manufacture of a small quantity of parts. The number of parts that is a small batch will depend upon the

A *Example of a one-off product: a luxury yacht*

B *Example of a product made in small batches: commercial aircraft*

type of product. For example, a small batch could be two engine parts for a Formula One car or 1000 resistors of a special value.

A large engineering company may have a wide range of manufacturing processes in its factory. Machines can be very expensive, so most or all of them will be in use making parts. Before they can be changed over to make different parts, there are a number of things that might need to be done:

- Moulds or work holding devices might need to be bought or made.
- Machinery may need to be programmed, which workers will have to be paid to do.
- Individual machines will need to be stopped, the tools changed and set-up to make the new part. During the set-up time, no parts are made, losing potential output.

If the machines are already in use making other products, the manufacturer might decide that they can make more money by continuing to make those products, rather than stopping their production just to make a small number of parts.

For this reason, small quantities of parts are often made either by hand or by using workshop equipment that can quickly be changed between different jobs. While the processes used may be of the same type, the equipment used may have significantly less automation and be controlled manually by a skilled worker.

Starter activities

1 Identify 10 products where the parts were probably made in small quantities.

2 A one-off part is sometimes made before making a large quantity of a product. Two of the aims of making this part can be to see how easy the product is to make and to assess the quality of the product. If the production machines are already in use making other products, what would be the advantages and disadvantages of having to use different equipment, operated by a skilled worker, to make these parts?

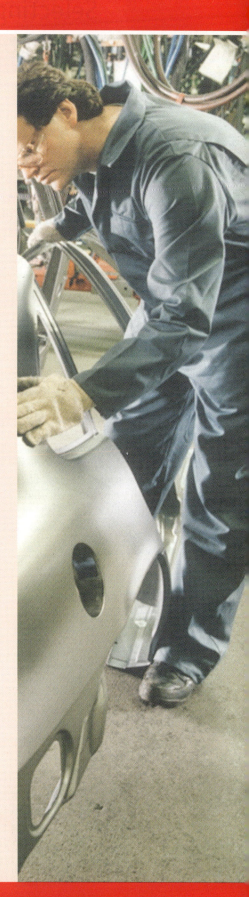

4.1 Selecting the processes to make the product

■ Types of process

Objectives

List five types of process.

Explain the type of operation carried out by each type of process.

Processes are operations that change the **form** of the materials. This means that they change the size, shape or condition of the material in some way. For each type of process, there may be a number of different types of equipment and tools that could carry it out.

The materials selected to make a part can be bought in a wide range of shapes and sizes, depending upon their type. Common forms range from powder, granules and liquids to sheets, slabs, bars and rods of various sizes, Diagram **A**. The designer or engineer will normally try to choose a form of material that requires the smallest possible amount of further processing to make the product.

Once the form of the materials that will be used to make a part has been decided, the next step is to determine the processes that will be used to change that form into the form of the finished part. For most parts, it will be necessary to carry out a number of different processes to achieve the required form. These processes can be broadly divided into the following types:

- materials removal
- shaping
- forming
- surface finishing
- joining and assembling.

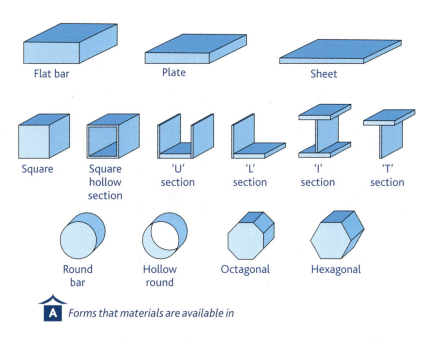

A *Forms that materials are available in*

In addition to the above, some materials may require 'conditioning'. Conditioning requirements are specific to individual materials. This includes things like heat treatment for some carbon steels or curing for composites, to allow the materials to achieve the required properties.

Materials removal

This involves using machines and tools to remove the parts of the material that are not needed in the required form of the part. It is sometimes known as 'wasting' or separation. There are a lot of different ways to carry out **materials removal**, including shearing, turning, milling, grinding, drilling and sawing.

B *Grinding – a materials removal process*

Shaping

Shaping involves pouring or forcing liquid material into moulds. This includes techniques such as casting and moulding. Once this material has become solid it is removed from the mould. This is the only process that can turn material into complex, three-dimensional forms in a single operation. Parts can often require some materials removal after shaping. For example, this can include material that has been squeezed into any gaps where the parts of a mould join together, which is known as 'flash'.

Forming

Forming changes the shape or size of the material, but normally does not affect the overall volume of the material. This is normally achieved by applying some form of force to the material, such as hammering, air pressure or a vacuum. Some materials are easier to form if they are heated prior to forming. However, unlike shaping, the material does not become liquid at any stage of the forming process.

Surface finishing

This involves modifying the surface of the part in a useful way. The aim of **finishing** is normally to improve one or more of the properties of the part, such as its appearance or its corrosion resistance. Finishing processes includes applying coatings, chemical treatments or polishing. The type of finishing process applied will depend upon the materials that the parts are made from and the properties needed. Finishing processes may be applied to individual or to assemblies of parts.

Joining and assembling

Joining and **assembling** involves attaching or putting the parts together. This might be achieved by bonding parts so that they are permanently attached or by using mechanical fastenings to hold them together. The joining process used will normally be specified in the design, as it can have a significant effect on the performance of the product.

C *Casting – a shaping process*

D *Welding – a joining process*

Activities

1. Using the internet, find a materials supplier or stockholder. For each material type, make a list of the different forms that are available.

2. Identify the different types of process (e.g. forming, shaping, joining – not the actual equipment or machine used) that would be needed to manufacture the following products:
   - The frame for a go-kart, made from steel tubes
   - The body of a kettle, made from a thermoplastic

Summary

Processes change the form of a material or part.

Common processes include material removal, shaping, forming, finishing, joining and assembly and conditioning.

A range of different processes may be needed to convert a material to the required form of the finished product

Key terms

Form: the size, shape and condition of a piece of material, product or part.

Materials removal: taking away the sections of the material not required in the final part.

Shaping: making a part by putting liquid material into a mould.

Forming: changing the size or shape of a material.

Finishing: modifying the surface of the part in a useful way.

Joining: attaching parts together.

Assembling: putting parts together.

AQA Examiner's tip

It is important to use a wide range of processes to demonstrate the range of your ability.

4.2 Materials removal (1)

Things to consider when selecting a tool

When you buy material to use to make a part, it is often bigger than the part you are going to make. This means that you have to take away the material that is not needed. Materials removal can be broadly divided into two types:

- Cutting, where the material has to be separated (or split) from a larger piece. Cutting operations include **sawing** and **shearing**.

- Materials removal, where excess material is removed to change the form of the part. There are many different ways to remove the excess material that is not needed by the part, including filing, **drilling**, turning, milling and grinding.

Choosing the most appropriate tool and method for cutting and material removal will depend on:

- the material being processed
- how much material needs to be cut off or removed
- the shape of the part being made
- the surface finish needed
- how accurate the part must be
- how big the part is and how easy it is to move it around
- the tools and equipment that are available
- the skill of the workers.

The engineer will also have to think about what form the removed material will be in and how it will be disposed off. It could be grinding dust, chips of material, swarf or large off-cuts. Where possible, this material is sent for recycling.

Methods of powering and controlling tools

When selecting a tool to remove material, the engineer also has to think about how it will be powered and controlled. Depending on the individual tool, materials removal and cutting can be carried out using hand tools, manual machine tools or computer controlled machine tools. The advantages and disadvantages of these approaches are compared in Table **A**.

A *Comparison of different methods of powering and controlling tools*

	Hand tools	Manually operated machine tools	Computer controlled machines
Equipment cost	Low	Moderate	High
Accuracy	Low	High	High
Skill to set up	Not needed	High	High
Skill to machine	High	High	Low
Machining time	High	Medium	Low

Objectives

Explain the things that need to be considered when selecting a tool for materials removal.

Compare different methods of powering and controlling tools.

Explain what is meant by CNC.

Compare the advantages and disadvantages of manual and computer controlled machining.

B *A manual mill*

Hand tools are designed to be held in the worker's hand. Depending upon the type of the tool, these might be powered by the muscle-power of the worker using the tool, electricity or compressed air. They are especially useful when the part being machined is difficult to move, for example when drilling holes for fixings into the hull of a ship.

C A computer controlled mill

Machine tools are usually fixed to the floor and attached to a power supply. This means that they can only be used when the part to be machined can be moved to them. They also need to be set up for the job before they are used. However they can be more accurate and work faster than hand tools, as well as being less tiring to the person doing the work! Manual machine tools need to be controlled by a skilled worker, who will guide the machine to make the part to the correct size and shape Photo **B**. **Computer Numerical Control (CNC)** machines use computers to control the machine, Photo **C**. All machines that are controlled by a computer are CNC because all machine code is made up of numbers.

The list of tools in Topic 4.3 includes vinyl cutters, laser cutters, abrasive water jet cutters and mills for making circuit boards. These machines all use 2-axis CNC control to move the tool. This means that the computer can move the tool around on a flat surface in any combination of two directions, called the *x* and *y* axes, Diagram **D**. This means that the tool can be moved backwards, forwards, right or left from its starting point, but not up or down. The advantages of using 2-axis CNC machines for making small numbers of parts are that they are faster and more accurate than manual machines, although they are very expensive to buy and they need to be programmed.

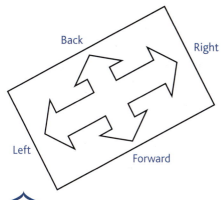

Back   Right

Left   Forward

D Axes of movement of a 2-D CNC machine

Activities

1 Use the internet to find examples of hand tools, manually operated machines and CNC machines, that could be used for the drilling of mild steel sheet. Compare the costs and features of the different types of equipment for drilling.

2 Find out how one of the 2-D CNC machines at your school is used. Design and make a simple shape using this machine. Make the same shape using hand tools. Compare the total time needed to make the two shapes and their accuracy. Explain any differences that you find.

Summary

When selecting tools for materials removal, the things you have to think about include the material, how much material needs to be removed, the accuracy and the equipment that can be used.

Materials removal can be carried out using hand tools, manual machine tools or computer controlled machine tools.

Computer Numerical Control (CNC) machines use computers to control the machine.

CNC machines can be faster and more accurate than manual machines and hand tools. However, they are very expensive to buy and they need to be programmed.

Key terms

Sawing: using a blade with sharp teeth to cut material.

Shearing: pushing a blade into a sheet of material to separate it.

Drilling: making holes.

Computer Numerical Control (CNC): using numerical data to control a machine.

AQA Examiner's tip

Use a 2-D CNC device in some of your practical assignment work to become familiar with the principles of operation.

▪ Types of tool

The aim of the following list is to show you the range of different tools that are available for material removal. It will not explain how each tool is used in detail, as this will probably be covered for the tools that you will have access to during your practical work in this subject.

Tools for cutting

Different types of saw have different sizes of teeth, to cut different types of material. The harder the material, the smaller the teeth are. Tenon saws have quite big teeth and are used to cut straight lines in wood. Coping saws have slightly smaller teeth and are used to cut curved lines in wood or plastic. Hacksaws have finer teeth and are used to cut plastic and metal. Mechanical saws are used to cut metal bars and rods. They are like large hacksaws, where the movement is powered by a motor.

Guillotines and shears can be used to cut metal sheet. These use a large force to press a blade made of tool steel through the material, in a shearing action.

Adhesive vinyl (sticky-back plastic) can be cut using a computer controlled knife in a vinyl cutter.

Parts can also be cut by melting along the line to be cut. For thermoplastic materials, this can be achieved using a heated wire. For ferrous metals, this might be done using the flame from an oxyacetylene torch.

Thin sheets of plastic or metal can be cut using a laser. The laser vaporises the material along the cut line.

Metals, plastics and ceramic materials can be cut by abrasive water jet cutting. This uses water containing sand, which is fired at the surface of the material at very high pressure. It wears away the material on the cut line.

Laser cutting and water jet cutting, Photo **A**, must be controlled by a computer. It would not be safe for people to use this equipment by hand. Although the equipment needed to carry out these methods is expensive, as these processes are very flexible they are often used for making small quantities of parts.

Tools for materials removal

Files are a hand tool. They have hundreds of small teeth to cut away at a material. Rough cut files are used to remove material and fine cut files are used for finishing. Most files are meant for metal and plastic. There are a lot of different shapes of file, Diagram **B**.

Drilling is used to make holes. This can be carried out manually using an eggbeater drill, a brace or a battery powered drill. If the part being drilled can be moved, you could use a pillar drill. Different types of drill bit are used for different materials. For example, high speed steel twist bits are used on metals and plastics.

A *Water jet cutting of aluminium*

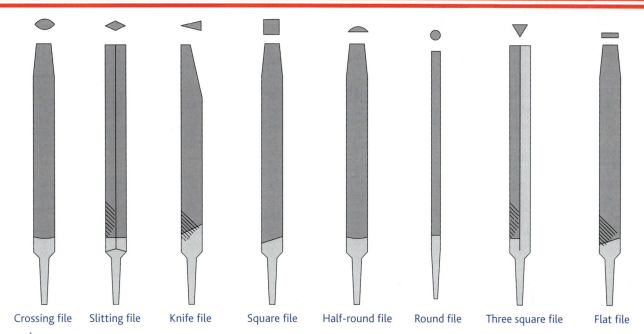

| Crossing file | Slitting file | Knife file | Square file | Half-round file | Round file | Three square file | Flat file |

 B *Types of file*

Turning is where cylindrical objects are made using a lathe. In a lathe, a piece of material is held by the machine and rotated at high speed. The tool or cutting bit is pressed into the work piece to remove material. Lathes can also be used to cut threads.

Milling machines are used to remove metal one thin layer at a time. They produce a flat surface with a good finish. Computer controlled mills can be used to make circuit boards, by **routing** away the material around the tracks between the components.

Grinding machines use spinning wheels made of abrasive materials. Compared with milling they remove less material each time they touch the part. They can be used to accurately shape a part and to produce a very smooth, mirror-like finish.

C *Manual milling machine*

Activities

1. Use the internet to find pictures of at least one example for each of the different methods of cutting and materials removal.

2. Identify the tools that you can use for cutting and materials removal in your school workshop.

3. Find out how one of the machine tools for materials removal is used. Write a detailed series of instructions so that someone who has not used that tool before would be able to use it safely.

Key terms

Turning: rotating a component at high speed while a cutting tool is pushed into it to remove material.

Milling: using a rotating tool to remove a thin layer of material.

Routing: a method of milling.

Grinding: using an abrasive wheel to remove a very thin layer of material.

Summary

Tools and equipment used for cutting include saws, guillotines, hot wire cutters, lasers and abrasive water jet cutters.

Tools and equipment used for materials removal include files, drills for making holes, lathes for turning, mills and grinders.

4.4 Shaping (1)

Shaping involves making parts by moulding liquid materials. Methods of shaping include the casting of metals, the injection moulding of plastics and the moulding of composites.

■ Casting

Casting is used to make three-dimensional shapes from metal. It is normally much cheaper and quicker to cast a complicated shape than to either machine it from a solid piece of metal or make it by joining lots of smaller parts together. During the process of casting, material is heated above its melting point so that it becomes liquid. The liquid is poured into a mould, and takes the shape of the hollow area in the mould. This hollow area is called the cavity. Once the material cools and becomes solid again, it is removed from the mould. Castings generally have good compressive strength.

Objectives

Understand why casting is used rather than other processes to make complex metal parts.

Learn about how casting is used to make parts.

Sand casting

Sand casting is used to make metal parts. It gets its name from using a mould that is made from a special type of sand.

The first thing you have to do when sand casting is to make a wooden version of the product that you want to make. This is called a **pattern**. The pattern should have no sharp corners, as these can be difficult to cast and may cause areas of high stress in the casting. The pattern is normally split into two halves, Diagram **A**.

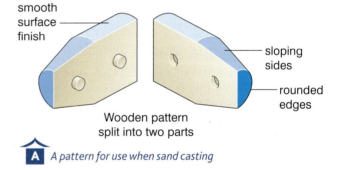

smooth surface finish

sloping sides

rounded edges

Wooden pattern split into two parts

A *A pattern for use when sand casting*

The bottom half of the pattern is put into a box known as a drag. Sand is poured in until the drag is full. The drag is then turned over and the pattern is taken out. In a similar way, the top part of the pattern is used to make a cavity in sand in a box known as a cope. The cope also includes a hole for the metal to be poured into (called a sprue), channels for the metal to move along (called runners) and a hole for any excess liquid metal to flow up into (called a riser). The drag and cope are assembled together to form the complete mould, Diagram **B**, and the liquid metal is then poured in, Photo **C**. This mould can only be used once – once the metal has set, the sand is knocked or shaken off the casting. The sprue, riser and any runners are then cut off the cast part.

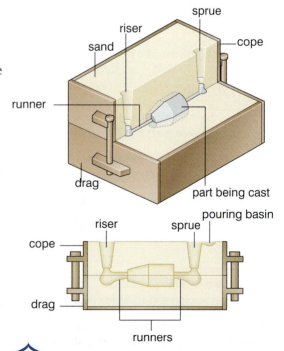

sprue

riser

sand

cope

runner

drag

part being cast

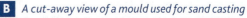

pouring basin

riser

sprue

cope

drag

runners

B *A cut-away view of a mould used for sand casting*

C *Sand casting*

Sand casting is used if you have to make a single part or a small batch of parts. If you have to make lots of parts of the same design, it would take a long time to make all the moulds from sand. Also, unless the sand moulds were lined up perfectly every time, the parts made might not be exactly the same.

Die casting

When lots of parts of the same design are to be made, you usually make a **die**. This is a special type of mould. It normally has two halves and is made from metal. Dies are often very expensive to make. They have to be made from metal with a higher melting point than the one which is being cast and be machined to a very smooth finish. Dies may also need channels machined in them to allow coolant to flow through the die.

During die casting, the parts of a die are brought together so that the liquid metal can be put into the cavity. Once the metal has set, the die is taken apart and the component removed.

In gravity die casting, the liquid metal is poured into the cavity using the force of gravity. In pressure die casting, the liquid metal is forced into the cavity under pressure. This can give more accurate castings than gravity die casting, with more detailed features, but needs more expensive equipment to inject the molten metal. The pressure die casting of metals is a similar approach to the injection moulding of plastics.

Castability

Some metals are less likely to have quality defects after casting than others. How well a metal can be cast is know as its castability. Diagram **D** and Table **E** show the castability and typical uses of a range of common metals.

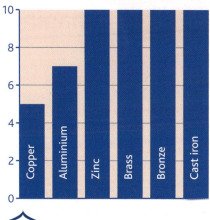

D　*Castability of common metals*

E　*Common uses of metal castings*

Metal	Castability	Example of what it is used for
Cast iron	Excellent	Engine blocks, engineering vices, anvils
Bronze	Excellent	Statues
Brass	Excellent	Door knobs
Zinc	Excellent	Model cars, camera bodies
Aluminium	Good/Excellent	Engine blocks for sports cars
Copper	Fair/Good	Bearings, bushes, valve bodies

Activities

1　Find a cast product. See if you can identify where the sprue, riser and any runners were attached.

2　Use the internet to find pictures of pressure die casting and products made using the different casting techniques.

Summary

Casting involves heating a metal above its melting point and pouring it into a mould.

It can be used to make complicated, three-dimensional metal shapes.

These shapes would be much more expensive if they had to be machined from solid metal or made by putting together lots of small parts.

Two common types of casting used to make metal products are sand casting and die casting.

4.5 Shaping (2)

Injection moulding

The **injection moulding** of thermoplastics is very versatile, producing items such as bowls, buckets, model construction kits, chairs and toys.

Diagram **A** shows how injection moulding is carried out. Plastic granules are loaded into a feed hopper. From here they drop into the barrel, where the rotating screw thread pushes the plastic along. It is melted by the heaters. At the end of the barrel the cone compresses the plastic and it is injected into the mould. The mould is cooled and the component is ejected. The mould can be moved back into position to make the next part within seconds.

B *Some products made by injection moulding*

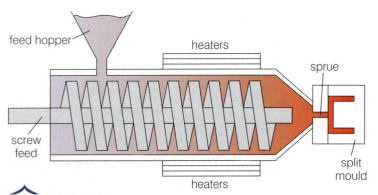

A *Schematic illustration of an injection moulding machine*

Although the equipment and moulds are very expensive, the process is very fast and complex shapes can be made. This means that normally it is only used when making thousands or tens of thousands of parts, so that the high cost can be divided between all of the parts made. However, it is sometimes used to make small quantities of expensive parts for special applications, such as spacecraft.

It is normally easy to identify an injection moulded part, as the sprue point where the plastic was injected is often visible, Diagram **C**. There may also be a 'split line' visible if the sides of the mould did not fit together perfectly.

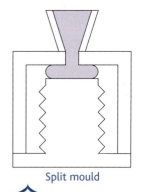

Split mould

Finished bottle top before sprue is cut off

C *Injection moulding of a bottle top, showing the sprue still attached*

Moulding composite parts

Fibreglass is a combination of flexible glass fibres and a hard and brittle thermosetting resin. When combined they produce a material that, weight for weight, is tougher than steel. It is used for canoes, boat hulls, car bodies, train bodies, furniture, baths, children's playground equipment and many other products.

Fibreglass products are easy to identify as they are smooth on one side and rough on the other. They are usually shaped by hand, using a one-piece **mould**. It is important that the mould is made the correct way so that the smooth side is where you want it (inside or outside depending on use), Diagram **D**. The mould is first covered in a gel-coat resin which gives the good finish. Layers of glass fibre are then placed in the mould and pushed into any corners. The glass fibres are then soaked in resin, which may be painted or stippled onto it. More layers of glass fibre are added and soaked with resin, until the desired thickness is reached. For high performance applications, the entire mould and part might then be covered with plastic sheets and a vacuum applied, to ensure that the resin has been pulled into all the spaces in the glass fibre. Depending upon the application and the resin used, the part may then be allowed to stand for 24 hours for the resin to go hard.

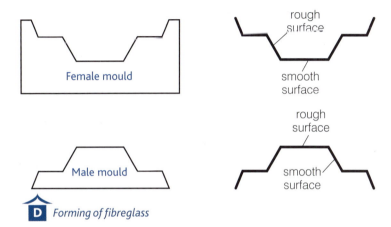

D *Forming of fibreglass*

This approach is very flexible and suitable for one-off or large batch production. The same approach is used to make parts from carbon-reinforced plastic (CRP). However, for CRP the carbon fibre material might be positioned so that more fibres run in one direction than another, to increase the properties in that direction.

E *Example of a composite product – the body of a hovercraft*

Activities

1. Find a plastic product that has been made by injection moulding. See if you can identify the sprue point.

2. Use the internet to find pictures of composite products being shaped using a mould.

3. Sintering is a method that can be used to shape some metals, ceramics and even composites. Carry out research to find out what this method involves and the sort of products that it can be used to make.

Summary

Injection moulding involves forcing liquid plastic into a mould. It can be used to make complicated, three-dimensional shapes.

The equipment and mould costs for injection moulding are very high, so it is normally only used when making either large quantities of parts or for high-cost parts for special applications.

Fibreglass and CRP are normally shaped in small quantities by hand, using a mould.

Forming involves using some form of force to change the size or shape of a material. There are cold and hot methods. Cold forming methods can only be used on malleable materials such as metal sheet, otherwise the material may break. Hot forming methods use heat to soften the material, making it easier to form.

Cold forming methods

Sheet bending

One way to bend sheet metal is to grip it in a folding bar and vice and hit it with a mallet or hammer. Hand operated presses and bending machines are also commonly available to make simple bends in thin sheet.

Press forming

Presses are commonly used to form metal sheet in industry. These presses are often powered by hydraulic rams, which can create massive pressures. One-off products and small batches can have bends made using simple angle fixtures. In mass production, automatic presses are used to form shapes such as car door panels using very expensive moulds made from tool steel, Photo **A**.

Hot forming methods

Historically, the earliest hot forming method was the hand forging of metal in a blacksmith's workshop. Hot forming methods used to manufacture small quantities of parts include strip bending and vacuum forming, both of which are used for thermoplastics.

Strip bending

This involves heating the thermoplastic sheet along a line, using a heating element. As the plastic heats it softens, allowing it to be bent. As it cools it will retain its shape. It is often done in schools using acrylic sheet, because it is a low cost process.

It is a good idea to make a wooden former to ensure accurate **bending**, Diagram **B**. It is important to allow the plastic to cool slightly before it is removed from the former – otherwise, if it is still hot it may sag, changing its shape.

Strip bending can be carried out by hand for one-off or small batch production. Automated machines can be used for the production of large batches.

A   *Example of a press formed part – car body panels*

wooden mould to ensure accurate bending of book stand

acrylic book stand

B   *Strip bending mould*

Vacuum forming

Vacuum forming is used to make many different products, including packaging, helmets, masks and baths. It uses sheets of thermoplastic. These are heated to make them flexible, formed over a mould, and then cooled to become hard again, Diagram **C**.

The moulds used are often made from wood or MDF and are much cheaper than those used for injection moulding. The sides of the mould must slope to allow the plastic product to be lifted off or pulled out, Diagram **D**. This slope is called the draft angle. It should be between 5 and 10 degrees. If there were no angle, the plastic product might remain stuck in the mould. The corners of the mould should have a radius. Any recesses must have small vent holes drilled in them to prevent trapped air stopping the plastic sheet forming.

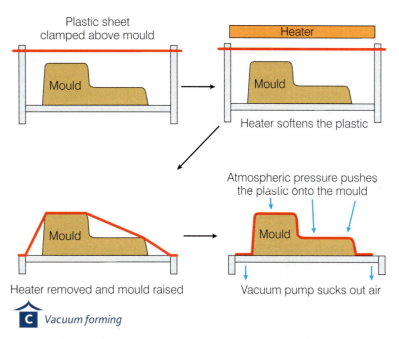

Plastic sheet clamped above mould

Heater

Mould

Mould

Heater softens the plastic

Atmospheric pressure pushes the plastic onto the mould

Mould

Mould

Heater removed and mould raised

Vacuum pump sucks out air

C *Vacuum forming*

Vacuum forming can only be used to make shapes of simple profiles, as any overlaps would cause the plastic to be stuck on to the mould. Unlike injection moulding, it cannot make reinforcing fins inside the shape.

Vacuum forming is commonly used for one-off and batch production. It is not often used for mass or continuous production as the shape has to be cut out from the plastic sheet. This adds an extra production step, and therefore extra cost, compared with processes that make the part using liquid plastic.

draft on sides — radiused corners

vent holes

D *Vacuum forming mould*

Activities

1. Find three examples of vacuum formed products. Identify the type of thermoplastic that may have been used to make each of them.

2. Two other hot forming methods are the drop forging of metals and compression forming for plastics. Use the internet to find out how these processes are carried out. Why are these methods normally only used when there is a large number of parts to make?

Key terms

Bending: forming an angle or curve in a single piece of material.

Vacuum forming: forming a thermoplastic sheet over a mould, using heat and a vacuum.

Summary

Cold forming uses direct force, such as a hammer or press, to change the shape of a metal sheet.

Hot forming methods, such as strip bending and vacuum forming, use heat to soften the material, making it easier to form.

Vacuum forming uses a cheaper mould than injection moulding, but is unable to make reinforcing fins within the shape. It is a slower process than injection moulding and normally requires more labour.

AQA Examiner's tip

You will need to be able to explain the differences between injection moulding and vacuum forming to make plastic products.

4.7 Joining and assembling

Joining means attaching parts together. There are two main types of joint, permanent and temporary. In permanent joints, either one of the parts being joined or the joining medium would have to be destroyed to take apart the assembly. Temporary joints can be taken apart and put back together again. Permanent joints are normally stronger than temporary joints which can loosen over time. However, temporary joints allow the product to be taken apart for maintenance and repairs to be carried out. The type of joint to use will have been decided during the design of the product, based on what the product is needed to do.

Assembling means putting the parts together. This may involve using joining techniques, but not every part in an assembly has to be attached to it. Some might be designed to allow movement or to be held together because they fit together tightly.

■ Permanent joining methods

Welding
Welding involves heating the area of the joint until the material melts and runs together. It can only be used when joining materials that are the same type. It often involves adding a filler material, to fill the gap in the joint. Welding processes for metals include manual metal arc (MMA) welding and metal inert gas (MIG) welding. These are carried out manually for small quantities of parts, Photo **A**. MIG welding can be automated for larger quantities. Welding processes for thermoplastics include hot plate welding, hot wire welding and friction welding.

Brazing and soldering
Brazing uses heat to join metals. However, unlike welding, the parts being joined are not melted, only the brazing material which flows into the joint. Soldering is very similar to brazing. It is likely that you will have used manual soldering at school. It is mainly used to attach components to circuit boards, Photo **B**.

Riveting
Riveted joints are used in applications such as attaching the skin to aircraft. The **rivets** are inserted into holes in overlapping pieces of material and the ends are made bigger to hold the materials in place, Diagram **C**. Rivets are mainly used to join metals.

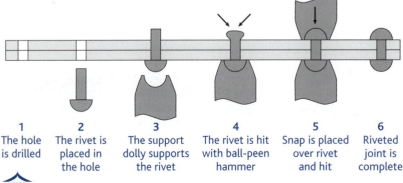

1	2	3	4	5	6
The hole is drilled	The rivet is placed in the hole	The support dolly supports the rivet	The rivet is hit with ball-peen hammer	Snap is placed over rivet and hit	Riveted joint is complete

 C Using rivets to join metal plates

Objectives

Explain the difference between permanent and temporary approaches to joining.

Describe a range of methods for joining and assembling.

A Manual welding

B Manual soldering

Key terms

Welding: a method where a joint is created by melting the contact areas of the parts to be joined.

Soldering: a method of joining electrical components to circuit boards.

Rivets: a type of mechanical fixing.

Adhesives: compounds used to chemically or physically bond items.

Threaded fastenings: mechanical parts such as screws, nuts and bolts.

Adhesives

There are a wide range of **adhesives** that are available to bond parts together. Different types of adhesive are used for different materials. Some can be used to join materials of different types. Three of the most common adhesives are:

- Epoxy resin: used for joining metals, plastics and composites. It is a type of polymer which comes in two parts, which harden when mixed together. It must be mixed immediately before use.
- Acrylic cement: joining acrylic and some other types of thermoplastic. This 'melts' the surface of the plastic and fuses it together.
- Polyvinyl acetate (PVA): a general purpose glue for woodwork.
- Most adhesive joints are weaker than the parts being joined. It is important that the surfaces to be joined are clean and free from dust and grease.

Temporary joining methods

It is possible to make temporary joints by crimping parts together, for example like twisting wires together in an electrical circuit. However, for mechanical applications temporary joints are most often made using **threaded fastenings**. These include nuts, bolts and screws, Diagram **D**. These are available in a wide range of materials, including steel, brass and different thermoplastics. With care, this type of fixing can be used to make joints in most materials, including different materials.

Assembling

For the manufacture of small quantities of parts, assembling might be carried out using simple hand tools, such as screwdrivers, spanners, mallets and hammers. If you are making more than one part, you could use powered tools, such as electric or pneumatic screwdrivers. These can work faster than hand powered tools, but mainly have the advantage that your hands don't normally get tired as fast when using them.

Activity

Carry out research to identify what joining methods you might use to join the following combinations of materials together.
- two pieces of mild steel plate, used for armour plating on a car
- an aluminium sheet onto a steel frame
- the fibreglass hull of a canoe to a steel mounting for a seat
- two different coloured pieces of a child's ride-on toy car, made from polypropylene.

Summary

Permanent methods of joining can create strong joints. Temporary methods of joining allow the product to be taken apart and reassembled.

Permanent methods of joining include welding, brazing, soldering, riveting and adhesives.

The most common temporary joining method is to use threaded fasteners, such as screws, nuts and bolts. These can be used to join dissimilar materials.

Hexagon head screw

Square head screw

Hexagon socket screw

Cylinder screw (pan-head type) slot

Cylinder screw, cross slot

Oval countersunk screw, slot

Oval countersunk screw, cross slot

Countersunk screw, slot

Countersunk screw, cross slot

Set screw slot

Wood and self-tapping screw

Wing screw

Hexagon nut

Crown nut

Square nut

Wing nut

D *Types of screw fastening*

Surface **finishing** involves modifying the surface of the part or product in a useful way. Many parts use some form of surface finish to improve their appearance, resistance to scratches or provide corrosion resistance.

Painting

Painting can be used to provide a decorative and corrosion resistant coating for metal surfaces. It is the easiest and cheapest way of coating an item. Paints consist of three components:

- Pigment – this is what provides the paint with its colour.
- Vehicle – this is the part that has to form a film when the paint dries and sticks to the surface.
- Solvent – this controls how easy the paint is to apply. It evaporates as the paint sets.

Paints can be applied by brushing, spraying or dipping.

Plating

Plating uses a process called electrolysis to coat the surface of a metal with a thin layer of material, Diagram **A**. This means that the part to be coated is put in a bath or tank containing a solution of chemical salts and an anode. The anode is a piece of the metal that will be used in the coating. An electric current is passed through the part, the solution and the anode. This causes a thin layer of metal to slowly build up on the part. At the same time, the anode gradually dissolves away. The amount of metal deposited depends on the concentration of the solution and the strength of the current. This approach is used to apply coatings of nickel, zinc, copper or tin.

Chromium plating is produced in a slightly different way. The anode does not dissolve into the chemical solution and extra salts have to be added to the solution instead. The part is normally nickel plated first, which improves the parts corrosion resistance, before the chromium plating is used to provide a decorative appearance.

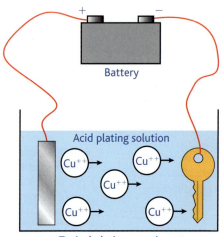

Typical plating experiment

A *Electroplating*

Key terms

Finishing: modifying the surface of the part in a useful way.

Painting: applying a liquid which dries to form a coating on the part.

Plating: depositing a layer of metal using electrolysis.

Anodising: the electrolysis of aluminium in an acidic solution.

Galvanising: dipping a steel part in molten zinc.

Etching: removal of some of the surface of a part by mechanical or chemical means.

Polishing: a physical process where the surface is rubbed or buffed to make it smoother.

AQA Examiner's tip

Be aware of a range of finishes and why they are used.

Anodising

Anodising is a plating process which is most often used on aluminium. It involves electrolysis in an acid solution. It provides a durable, corrosion-resistant finish. Colour can be added to tint the aluminium for decorative purposes. Anodising is also used to make the dielectric films that are used in electrolytic capacitors. Anodised titanium is used for dental implants.

Galvanising

Galvanising involves dipping metal into a bath of molten zinc. The liquid zinc sticks to the steel and cools to form the coating. It is normally used for mild steel. The zinc provides good corrosion resistance, although the appearance of the coated part is not attractive. Galvanising is used to coat the materials used to make metal dustbins, buckets, animal feeding troughs and road signs (Photo **B**).

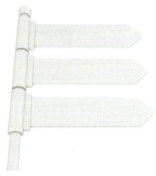

B *Example of a galvanised product – road sign*

Etching

During **etching**, selected areas of the surface are dissolved by chemical action. It is commonly used in the production of printed circuit boards (PCBs), where the tracks needed for the electricity to pass through when the circuit is made are protected from the etching solution (Photo **C**). Etching is also used to provide decorative finishes on metals and on ceramic items, such as glass. The areas that the chemicals can touch and the amount of time that the chemicals can contact the part must be carefully controlled to get the desired effect.

Polishing

Polishing is a physical process which gives the surface a smoother, more shiny appearance. As well as giving a decorative finish, this can help to make the surface more resistant to fatigue failure and to reduce friction. It involves removing a tiny amount of the material on the surface of the part. This can be carried out with a non-abrasive cloth or a buffing wheel.

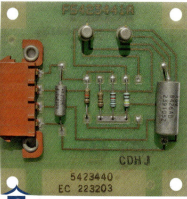

C *The copper tracks between the components on a PCB are produced by etching*

Summary

Surface finishing can be used to improve the appearance, resistance to scratches or corrosion resistance of a part.

Plating uses electrolysis to apply a thin layer of material on to the surface of a metal part.

Coatings can also be applied by painting or dipping in molten material.

Chemical action and polishing can also be used to change the surface of a part.

Activities

1. Look around your home and identify at least six products that have been given some form of coating. Explain the benefit that the coating or surface finish brings to each of these parts.

2. Two other surface finishing methods that are used in industry are plastic dip-coating and powder coating. Use the internet to find out how these methods are carried out and to identify products that they are used on.

Selecting the materials to make the product

Chooser charts

There are a large number of different processes and equipment. Many of these are suitable only for certain types of material or forms of material. The engineers have to look at the design for a product and the materials they have to make it from, then decide which processes to use to make the product.

The engineers may use a chooser chart, like that shown in Table **A**, as a starting point. However, there are a large number of different things to consider when selecting the best tool. They will need to use their knowledge and experience or to test the processes to make sure that they can achieve the required task.

Objectives

Look at the comparative properties of different materials.

Explain how a chooser chart can be used to assist the selection of materials for an application.

A *Methods used to manufacture different materials*

Process	Tool or Machine:	Mild steel sheet	Mild steel bar	Aluminium alloy	Melamine	Acrylic sheet	Polypropylene	Alumina	Fibreglass (GRP)
	Material type	Ferrous material	Ferrous material	Non-ferrous material	Thermoset	Thermo-plastic	Thermo-plastic	Ceramic	Composite
Material removal	File	Yes	Yes	Yes	Maybe	Yes	Maybe	No	Maybe
	Drill	Yes	Yes	Yes	Yes	Yes	Yes	No	Yes
	Engineer's lathe	No	Yes	Yes	No	Yes	No	No	No
	Milling machine	Yes	Yes	Yes	No	Yes	No	No	No
	Grinding machine	Yes	Yes	Yes	No	No	No	No	No
Cutting	Hacksaw	Yes	Yes	Yes	Yes	Yes	Yes	No	Maybe
	Mechanical saw	Maybe	Yes	Maybe	No	No	No	No	No
	Guillotine/shears	Yes	Maybe	Yes	Maybe	No	No	No	No
	Oxyacetylene cutting	Yes	Yes	No	No	No	No	No	No
	Laser cutting	No	No	No	No	Yes	Yes	No	No
	Abrasive water jet cutting	Yes	Yes	Yes	No	Maybe	Maybe	No	No
Shaping	Sand casting	N/A	N/A	Maybe	No	No	No	No	No
	Pressure die casting	N/A	N/A	Yes	No	No	No	No	No
	Injection moulding	No	No	No	No	Yes	Yes	No	No
	Moulding	No	No	No	No	No	No	No	Yes
	Sintering	No	No	No	No	No	No	Yes	No
Forming	Hammer	Yes	Yes	Yes	No	No	No	No	No
	Press	Yes	Yes	Yes	No	No	No	No	No
	Strip bending	No	No	No	No	Yes	Yes	No	No
	Vacuum former	No	No	No	No	Yes	Yes	No	No
	Forging	No	Yes	No	No	No	No	No	No
	Compression moulding	No	No	No	Yes	No	No	No	No

Joining	Welding and brazing	Yes	Yes	Yes	No	Maybe	Maybe	No	No
	Adhesives	Maybe	Maybe	No	Maybe	Yes	Yes	No	Yes
	Riveting	Yes	Maybe	Yes	No	No	No	No	No
	Screwed fasteners	Yes	Yes	Yes	Yes	Yes	Yes	Yes	Yes
Surface finishing	Painting	Yes	Yes	Maybe	No	No	No	No	Yes
	Plating	Yes	Yes	Yes	No	No	No	No	No
	Anodising	No	No	Yes	No	No	No	No	No
	Galvanising	Yes	Yes	No	No	No	No	No	No
	Etching	Yes	Yes	Yes	No	Yes	No	Yes	No
	Polishing	Yes	Yes	Yes	No	Maybe	No	No	No

Case study

Process selection: back plate for the emergency wind generator

The aim of this case study is to show how an engineer would look at the design of a component for the emergency wind power generator developed in Chapter 2 and decide which processes and equipment could be used to make it.

The back plates of the body, Photo **B**, are made of aluminium alloy. The edges of the plates are curved over. There are several holes, in a number of different sizes. The aluminium alloy is available in sheet form, much larger than required.

To cut the sheet to the size needed, a hacksaw, guillotine or abrasive water jet cutting could be used.

To make the holes a drill could be used. This could be a hand-held electric drill or a pillar drill.

To make the sheet into the required shape, it could be hammered by hand or a press could be used.

The engineer will consider each of the process choices, taking into account the number of parts to be made, the accuracy required and the equipment available. To make a one-off in a typical school workshop, he would probably select a hand operated guillotine to cut the sheet, as this would be more accurate than a hacksaw and it is unlikely that he would have access to an abrasive water jet cutter. He would probably use a pillar drill to make the holes as, with work held in a vice, this may be more accurate than a hand-held drill. Unless he had a special press for bending, he would probably use a hammer and vice to make the bend in the plate. The sequence of processes to use will be discussed in Topic 4.10 on production planning.

B   *Back plate of the wind power generator*

Activity

For each of the following products, identify the processes and equipment that could be used to make them.
- the frame of a cheap mountain bike, made from steel tubes
- the body of a kettle, made from a thermoplastic
- the composite shell of a canoe.

Summary

There is a wide range of equipment available to carry out different manufacturing processes. Each tool has different characteristics. Some equipment may be suitable only for a small range of materials.

A chooser chart is a useful tool to help select the tools and equipment that could be used to make a product. However, the engineer will still have to consider the number of parts to be made, the accuracy needed and the equipment that is available before making the final choice.

AQA *Examiner's tip*

You need to be aware of why each process is used as well as how it is carried out.

4.10 | The production plan

The role of the production plan

A **production plan** is a set of instructions for making the part. It should contain enough information that someone who has never seen the finished part should be able to make it. The main reasons for preparing a production plan are shown in Table **A**.

A *Reasons for preparing a production plan*

Type of benefit	Reason
Quality control	It helps to prevent design features being missed during manufacture
Consistency	If the same product has to be made again in the future, it can be made the same way
Safe working	The production planner has to consider any risks that might arise during the manufacturing processes and will take action to reduce them. It also helps the people making the part be aware of what they need to do to work safely
Efficient use of machines	It helps the company to plan the use of machines. This can be very important in a busy workshop where there are a lot of different parts that need to be made on the same machine
Financial management	Material needs can be planned, so that they can be bought only for when they are needed, rather than having to take up expensive storage space

Putting the processes in order

The first task for the engineer is to look at the design, see what materials are available to make it and to identify which processes need to be carried out to turn the materials into the finished part. For example, consider the back plate of the wind power generator designed in Chapter 2. If a large sheet of aluminium were available to make it, the engineer may identify that the following processes are needed.

- cutting, to reduce the sheet to the size needed
- forming, to bend the sides
- materials removal, to make the holes.

In addition, the engineer would identify that the part would need marking out for each of these activities.

Next, the tasks need to be put in a suitable order. Some tasks will need to be carried out before others can begin. For example, for the back plate, the holes will need to be drilled before the sides are bent as otherwise the part may not fit in the machine vice on the drill. The engineer normally carries out the **sequencing** by using his or her knowledge of the manufacturing processes. The final sequence can be shown as a **flowchart**, Chart **B**.

Creating the production plan

On its own, the sequence of tasks does not give enough information to make the product. The other things that the person making the product needs to know before they begin are detailed in the production plan, Table **C**.

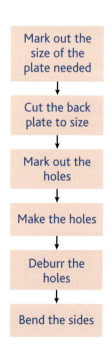

B *Flowchart showing the sequence of tasks needed to make the back plate*

As part of this activity, the engineer should identify exactly which tools are to be used. He or she should also explain what alternatives could be used if these are not available, and any impact that these may have on the manufacture of the product. The engineer would also carry out a risk analysis for each piece of tool or equipment that could be used – this will be explained in the next topic.

C *Extract from a production plan for the manufacture of the back plate of the emergency wind power generator*

Step	Task	Links to specification	Time, minutes	Tools to use	Materials to use	Quality control	Safety notes
1	Mark out the size of the plate needed and the lines to show where the plate should be bent	11, 13, 14	15	Scribe, engineers blue, engineering rule	Aluminium sheet, 1000 × 1000 × 1.2 mm	Check the marked out dimensions using an engineering rule	Wear gloves when handling the sheet, as it may have sharp edges
2	Cut the plate to the size needed	12, 14	5	Guillotine	Aluminium back plate 125 × 65 × 1.2 mm	Check the dimensions using an engineering rule	Use the machine guard. Wear gloves when handling the large sheet, as it may have sharp edges
3	Mark out and centre punch the centre points of the holes to be drilled. See working drawing for dimensions	1, 2, 3, 14	30	Scribe, engineers blue, centre punch	Back plate from task 2	Check that the holes are marked in the correct positions using the template. If wrong, repeat task	
4	Make the holes in the plate. See working drawing for the drill bit sizes to use	1, 2, 14	20	Pillar drill, HSS drill bits	Back plate from task 3	Check the hole diameters using vernier calipers. Check the hole positions (distance to the edges) using an engineering rule. If the holes are too small, re-drill with the correct drill bit. If the holes are too big or in the wrong position, go back to step 1	Use the machine guard and wear goggles. Use a machine vice to hold the plate securely

Activity

List all of the activities involved with making a cup of coffee. Create a flowchart showing the correct sequence of the activities. Create a production plan that could be used by someone who had never seen a cup of coffee being made.

Summary

The production plan provides detailed instructions on how to make the product.

It should include all the tasks to be carried out, in the correct order.

It should also include all of the other information needed to make the part, such as the materials and tools to be used, any quality checks to be carried out and safety notes.

Key terms

Production plan: the instructions on how to manufacture a product.

Sequencing: putting a series of events in order.

Flowchart: a diagram showing a sequence of operations

AQA *Examiner's tip*

Break the processes into small operations and make sure that the sequences make sense.

Ensure your plans have sufficient detail for a third party to be able to use them.

What is a risk assessment?

Risk assessments identify any potential hazards associated with an activity. This means that they list any ways that the activity could cause harm. Further, the risk assessment also lists the risk mitigation actions. These are the things that need to be done to make sure that any risks identified are made as small as possible.

Risk assessments are carried out for a wide range of activities, ranging from sports to business operations to travel. Here we will focus only on their use during manufacturing operations involving the use of tools or equipment.

The aim of the risk assessment in manufacturing is to reduce the risk of harm to:

- the people carrying out the activity, such as the process operators
- other people who may be affected, such as nearby workers, visitors and managers
- the working environment.

Risk assessments should be prepared for every activity carried out during the manufacture of a product. These should be practical documents that are used by the people carrying out the operations to help make sure that they work safely.

How to carry out a risk assessment

The first step in carrying out a risk assessment is to identify possible sources of harm. For example, during a machining operation, the following questions might be asked.

- Are there rotating or moving parts that the operator could get caught in?
- Does the process produce any debris? How would this affect the operator, nearby people and the environment?
- How noisy is the process?
- What happens if the tool breaks or shatters during the process?
- Are there any features of the work piece that could cause harm?
- Is the work piece fixed in place during the operation or could it become a **hazard** if it moves?
- Is the process or work piece at a temperature that could cause harm?

Each of these risks should then be rated, by their potential impact and how likely they are to occur. One way of doing this is to use a high, medium and low rating system, as shown in Table **B**. For example, a high impact shows that the risk involved in manual soldering, Photo **C**, could cause a lot of harm, whereas a low impact shows that any resulting injury would be minor. It is important to remember that this type of rating is subjective and will vary between different applications.

 A Examples of some hazards and mitigation actions during a drilling operation

Hazard	Mitigation actions
Machine has moving parts	Use machine guard. Tie back hair. Secure any loose clothing, such as ties
Machining operation produces swarf	Wear safety goggles. Wear an apron
Tool may break or shatter during the process	Use machine guard. Wear safety goggles
Work piece has sharp edges	Wear gloves
Work piece is heavy	Use correct lifting technique. Lift with someone else. Use a crane to move it
Work piece might move during process	Use a machine vice or clamp to hold it in place
Work piece becomes hot during the process	Allow it to cool down before moving it

B *Example of a risk assessment for soldering*

Risk	Who this affects	Impact	Likelihood	Mitigation actions
1. The tip of the soldering iron is very hot. This can cause burns to hands and scorch other objects it touches	Operator, equipment used in the working environment	Medium	High	Always assume that the soldering iron is hot – never touch the hot end or you might get burnt. Never touch anything with the hot end except for what you are soldering. Always use a baseboard, so that the work surface does not get burnt. Always use a stand for the soldering iron, so that the work surface does not get burnt
2. Heat can also be conducted down the solder wire, causing burns	Operator	Low	Medium	Never use a piece of solder less than 5 cm long. This stops your fingers getting burnt by heat passing down the wire
3. Soldering makes smoke called soldering fume. This is poisonous and can make you ill	Operator, nearby workers, the environment	High	Low	Use a fume extractor to take away the fume

Risk mitigation actions

Once the risks have been identified, the next step is to identify how each of them can be reduced. There may be one or a number of actions for each risk. There is a broad range of possible **mitigation actions**, which include:

- using equipment such as machine guards or fume extractors
- following a written safe working procedure to use the process or having specialist training to use it
- using appropriate Personal Protective Equipment (PPE), such as goggles, gloves, ear defenders, breathing masks or aprons.

Table **A** on the previous page shows examples of mitigation actions that are commonly used during the drilling of sheet metal.

C *Manual soldering*

Activities

1. Create a risk assessment for one of the following tools that you are familiar with: drill, lathe or mill.

2. Create a risk assessment for one of the following processes that you have not yet used: vacuum forming, sand casting or abrasive water jet cutting.

3. Using the internet, find pictures of guards used in at least five different types of machine.

Key terms

Risk assessment: a review of the potential of an activity to cause harm.

Hazard: something that causes a risk of harm or injury.

Mitigation action: a precaution taken to reduce a hazard.

Summary

A risk assessment evaluates any potential hazards associated with the use of a tool or a piece of equipment.

The risk assessment should include mitigation actions to reduce the risks. These could include the use of machine guards, personal protective equipment and safe working procedures.

AQA *Examiner's tip*

It is essential that you can work safely with any process that you need to use.

4.12 Preparing for production

Quality assurance

An important part of preparing for production is **quality assurance**. This is about planning and doing things to make sure that the part being made satisfies the needs of the design.

Quality assurance is not the same as quality control. Quality assurance is a preventative activity. It is carried out before and during manufacturing activities. It is about making sure that the product is manufactured correctly. Quality control is a reactive activity. It is carried out after individual machining activities. It aims to find products that were not made correctly.

Quality assurance and quality control are both important during manufacturing. However, it is obviously cheaper and quicker to make the product right in the first place rather than have to correct it or make it again.

Templates, jigs and fixtures

An important quality assurance activity is preparing simple devices that will help to make sure that the part will be made to the sizes on the drawing. The most common types of device are templates, jigs, and fixtures. While these devices can help to ensure that one-offs or prototypes are made right the first time, this has to be balanced against the time needed to make them. However, they have many advantages when making small quantities of parts:

- They reduce the amount of time needed to set-up machining operations, by reducing the marking out required.
- They ensure that all of the parts are made the same and reduce scrap or rework.

Templates

A **template** is a standard shape that is used as a pattern. Templates can be made from paper, card, plastic or metal, depending upon how often they will be used and how long they are needed to last. They are used to:

- speed up the marking out of a one-off part with a complicated shape, for example a custom-made guitar
- mark out a series of identically shaped pieces
- ensure that a series of drill holes are always located in the same place on a repeat part, as shown in Diagram **A**.
- If the drawings for a part have been produced using Computer Aided Design (CAD), it is often possible to produce the template directly from the CAD drawing. This can be printed using either a normal printer or sent to a computer-controlled machine, if the template is made from another material.

Templates can also be used for quality control purposes, such as checking that the shape of a part is correct. For example, a paper or hardboard template could be used to ensure that four turned parts are all the same, Diagram **B**.

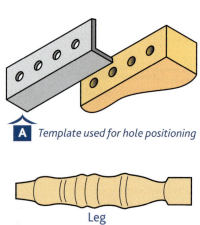

A *Template used for hole positioning*

Leg

Template

B *Template used to check the shape of a table leg*

Jigs

Jigs are made to help, hold or position a part, to achieve a consistent, repeated end result. If you are doing an operation once it may not be worth making a jig. However, if the process is repeated a jig can save a lot of time and effort.

For example, if three holes have to be drilled accurately and consistently in 50 pieces of steel, then a jig would be designed that would hold the work piece in the same position each time and guide the drill bit so that every hole was in the right place. Diagram **C** shows a suitable jig. The steel plate is clamped into a holder by the use of a cam. The drill bit would be guided by hardened steel bushes. These are used to prevent wear. They can be replaced, if necessary, to extend the life of the jig.

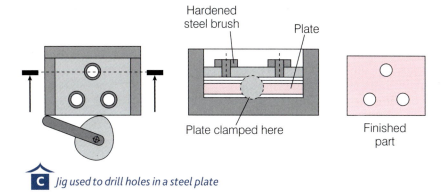

Hardened steel brush

Plate

Plate clamped here

Finished part

C *Jig used to drill holes in a steel plate*

Fixtures

Like a jig, a **fixture** is also a type of work holding or positioning device. A fixture is different from a jig only in that it is normally fixed to the bed of the machine tool, whereas a jig is moveable in order to line up with the tooling.

Activities

1. Use the internet to find a range of examples of different types of template, jig and fixture.

2. Design a jig that could be used to drill a 6 mm diameter hole through a 25 mm thick piece of square-section steel. The hole should be positioned centrally between the two sides and at an angle of 45° to the surface.

Summary

Quality assurance is a preventative activity concerned with ensuring that the product is made right. Quality control is a reactive activity to find products that were not made correctly.

A template can be used as a pattern to mark out parts or to check that they are accurate.

A jig is a removable device used to help hold or position a part. A fixture is a similar work holding or positioning device that is attached to a machine base.

Key terms

Quality assurance: planning and doing things to make sure that the product is made right the first time.

Templates: a standard shape that is used as a pattern.

Jig: a work holding or positioning device that is not fixed to the machine bed.

Fixture: a work holding or positioning device that is fixed to the machine bed.

AQA Examiner's tip

Use jigs, fixtures and templates to ensure that your work is accurately produced.

The practical work that you will carry out forms an important part of your assessment in GCSE Engineering. You will be assessed both on how well you work and on the standard of work you produce.

Working safely

Workshops have the potential to be very dangerous places. Every year there are a large number of serious accidents in manufacturing companies. People suffer broken bones, lose fingers or limbs, or even die, because they are not working safely.

It is a common mistake to think that the employer or school has sole responsibility for safety in the workshop. They do have to meet the requirements of many laws to provide a safe working environment. However, the Health and Safety at Work Act makes both the employer and the employee (you) equally responsible for safe working. Both can be prosecuted for not following safety regulations and procedures. Further, you can also be prosecuted if you interfere with equipment provided for your health and safety or the health and safety of others.

The following general guidance will help you to work safely. There will also be requirements unique to the workshop that you will work in and for the equipment that you will use.

Make sure that you know how to carry out any activities that you need to do safely. This means that for each activity you will carry out, you need to use all the mitigation actions shown on the **risk assessment**. These may include, for example, wearing appropriate Personal Protective Equipment (PPE), Photo **A**, using machine guards and following safe working procedures.

Do not use any equipment that is damaged or incomplete. For example, don't use electrical equipment if the wires to it are frayed or damaged; don't use a machine if its guard is cracked. Tell whoever is responsible for supervising the practical work of the problem immediately.

Don't lift heavy items by yourself. Ask for help or use lifting equipment, if needed.

Always clean away any mess that you make. This includes returning any tools and materials to the correct places.

A *PPE*

Information

Some common causes of accidents in the workshop

- Carelessness or loss of concentration
- Not following safety rules
- Lack of training in using the equipment
- Unguarded or badly maintained equipment
- Tiredness and fatigue
- Slippery floors and messy work areas
- Poor behaviour – not working sensibly
- Trying to lift items that are too heavy

Be aware of what to do in case of an emergency. For example, check out the fire procedures and be aware of the locations of the different types of fire extinguisher; know where the emergency stop buttons are; find out what to do if there is an accident.

■ Making the product

To make a product that meets the requirements of the design, you need to correctly interpret the working drawing and use your practical skills to achieve the stated dimensions and finish.

You should find the product dimensions directly from the working drawing(s). When using these, make sure that you understand the **conventions** used. These were explained in Chapter 2. One of the most common mistakes when using working drawings is to measure a dimension straight from the drawing, without taking into account the **scale** of the drawing.

The practical activities that you must carry out to achieve the design should be listed in your production plan. If there are any activities or processes where you are not confident that you can manufacture parts to the necessary standard or dimensions, it may be a good idea to carry out a practice activity using scrap material first.

It can sometimes be necessary to vary from the production plan. For example, this may be because the planned process was not available, or because it was necessary to correct a quality problem. You should keep a record of when this happens, along with an explanation of why the change was necessary.

To demonstrate that your part meets the requirements of the drawing, it can be useful to maintain a log of all the measurements carried out as part of the quality control. This will be explained in the next topic.

Activities

1 Using the internet, find out about the Health and Safety at Work Act. Prepare a presentation summarising:
   - the responsibilities of the employer
   - the responsibilities of the employee
   - the role of the Health and Safety Executive
   - who can be prosecuted under the act, and what the maximum penalties are.

2 Produce an A3 poster that could be used in a school workshop to explain safe working.

Key terms

Risk assessment: a review of the potential of an activity to cause harm.

Conventions: rules of presentation that drawings must conform to.

Scale: the ratio of the size of the design in the drawing to the size of the part or finished product.

Summary

The coursework assessment includes both how well you work and on the standard of work you produce.

The employee and the employee (you) are equally responsible for safe working.

Making a product that meets the requirements of the design needs correct interpretation of the working drawing and practical skills in the processes carried out.

AQA Examiner's tip

Remember that you are being observed during practical work. The way in which you work will affect your grade and future employment.

Making sure that the part is the correct size is vital if the product is to perform as it was designed. When making a one-off or a small quantity of parts, the most important activity towards achieving the correct **dimensions** is normally marking out. It is also important to check that the parts are the size intended after manufacturing. This is the main function of quality control.

Objectives

Explain why marking out is important.

List a range of tools that may be used for marking out.

Describe three measurement tools that are used for quality control.

Marking out

Marking out is the stage where the design is drawn onto the materials that the part will be made from. It cannot be over-emphasised how important correct marking out is. If the material is not marked out correctly, it does not matter how accurate a machining process is – the final product will not meet the requirements.

There are two types of line which need to be marked out: cutting lines, which show where parts are to be cut, and construction lines, which are used for all the other features. Different tools are used to make the two types of line in different materials, Table **A**. In addition, a centre punch may be used to mark out the centre of holes in metals.

Marking out should always be carried out from a true edge or specified point, which is called a **datum**. A range of different tools can be used to work out where the lines should be marked, depending on exactly what is required, Table **B**. For example, odd leg callipers are used to mark a line parallel to an edge on plastic or metal, and tri-squares and engineer's squares are used to mark lines at right angles to an edge.

Quality control

The aim of **quality control** is to find parts that were not made to the correct dimensions. As explained in Topic 4.12, it is a reactive activity. It normally involves measuring parts after individual machining activities. There is a wide range of different measuring tools available. Choosing the right tool depends on:

- the type of measurement that needs to be made
- the accuracy of measurement needed
- how repeatable these measurements are
- the size range that needs to be measured
- the cost of the tool
- how easy they are to use
- how difficult they are to **calibrate**.

In general, the more accurate a measuring device the more expensive it is and the harder it is to use. Three of the most common tools to measure dimensions are engineer's rules, vernier callipers and micrometers.

A *Tools for marking out*

Tool to use	Cutting line	Construction line
Wood	Marking knife	Pencil
Plastic	Scriber	Felt tip or wax crayon
Metal	Dot punch, making dots at 5 mm intervals	Scriber

B *Tools used for marking out*

Tool	Wood	Plastic	Metal
Angle plate		Yes	Yes
Centre punch		Yes	Yes
Compass	Yes	Yes	
Dot punch		Yes	Yes
Engineer's rule	Yes	Yes	Yes
Engineer's square		Yes	Yes
Marking gauge	Yes	Yes	
Mitre square	Yes	Yes	
Odd leg callipers		Yes	Yes
Sliding bevel	Yes	Yes	Yes
Surface plate		Yes	Yes
Templates	Yes	Yes	Yes
Tri-square	Yes	Yes	
Vee block		Yes	Yes

Engineering rules are used to measure straight lengths. They are normally made of steel and are relatively cheap. They come in a range of sizes, from 150 mm to 1 m long. They can be accurate to 0.5 mm with careful use. The measurement scale starts from the end of the rule. They are designed to be pushed up to the shoulder of a component or lined up with the end.

Vernier callipers can be used to make a wide range of measurements, including lengths, outside diameters, inside diameters (such as hole sizes) and hole depths, Photo **C**. A range of sizes are available, for measurements up to 600 mm long. They can be accurate to 0.1 mm, depending upon the ability of the user.

Micrometers are used to measure outside diameters and lengths, Photo **D**. They are more accurate than vernier callipers. Individual micrometers can only measure a range of 25 mm, so many applications need a range of micrometers in 25 mm steps (0–25 mm, 25–50 mm up to 125–150 mm). Different designs of micrometers are also available to measure hole depths and inside diameters.

C   *Vernier callipers*

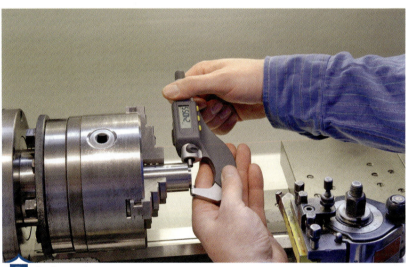

D   *A micrometer*

> **Key terms**
>
> **Dimensions:** the sizes of the parts.
>
> **Quality control:** checking that the part is correct after it has been made.
>
> **Calibrate:** checking, and if necessary modifying, a measuring tool to make sure it is measuring accurately.

> **AQA** *Examiner's tip*
>
> Measure twice, cut once – after you have finished marking out a part, check every measurement again to make sure that they are all correct. This is a lot quicker than having to change a part if it is wrong.

Activities

1 Mark out a template for a steel panel on a piece of card. It should be 100 mm square with a radius of 5 mm at each corner. It should have four holes, one in each corner, each of diameter 10 mm, each located 15 mm from the two edges.

2 Find a nut and bolt. Using a variety of measurement devices, measure all of the dimensions and produce working drawings of the two parts.

Summary

Marking out is the most important activity towards achieving the correct dimensions of a part. There are a wide range of tools available to support the marking out of different materials and different types of dimension.

Quality control is a reactive activity to find products that were not made to the correct dimensions. The measurement tools used for quality control include engineering rules, vernier callipers and micrometers.

4.15 Evaluation of the finished product

The reason for functional testing

As a result of the quality checking and measurements during production, the designer will hopefully have a list showing that every dimension on the drawing is within the stated tolerance. However, this does not guarantee that the product will work as required. For example, there may have been an error during the design or making; or it could be that the combination of tolerances for different parts means that moving parts stick together or are too loose. For this reason, it is essential to carry out **functional testing**. This means testing the product in a way that shows how it will work when it is used by the customer. This might mean trying it out. This could also mean running tests – for example, some manufacturers of chairs have a machine which simulates a large person sitting on the chair 10 000 times or more, to check that it does not fail over that time.

> ### Objectives
>
> Explain why it is important to carry out functional testing of the finished product.
>
> Explain why it is important to compare the product with the specification.
>
> Explain how the evaluation could be used to further improve the product.

Comparing the product to the specification

While functional testing checks that the product works as intended, it may not cover all of the needs that the product must satisfy, such as cost, aesthetic appeal or recyclability. To ensure that all the relevant needs are evaluated it is necessary to test every need in the **specification**.

This will probably mean that a number of different tests need to be carried out. Table **A** shows an evaluation against the specification for the wind power generator designed in Chapter 2. (See Photo **B**.)

Where possible, the testing should be objective. This means that it should be based on facts and numbers, rather than subjective opinions. The results of this testing are often shared with the clients, so that they can see how well the product meets their needs.

A *Example of an evaluation against the specification for the emergency wind power generator*

No.	Need	How tested	Result of testing	Outcome
1	The wind turbine shall be able to generate at least 100 milliamps at 12 volts in a wind speed of 5 metres per second.	Ammeter/voltmeter and a fan at a wind speed of 5 metres per second	Current output of between 150 and 250 milliamps at a voltage of 12 volts ±10%	Pass
2	The wind turbine shall be robust enough to operate in a wind speed of up to 5 metres per second for at least one hour.	The turbine was tested in a 5-metre wind speed for a period of 2 hours	There were no changes to the structure and current output was the same as the start	Pass
3	The wind turbine shall be mounted on a rod, which is between 7 mm and 9 mm in diameter.	Measure with vernier callipers	8.01 mm	Pass
4	The mounting rod shall have a length of at least 80 mm.	Measure with an engineering rule	110 mm	Pass
5	The blades shall rotate within an operating envelope of less than 400 mm.	Measure with an engineering rule	393 mm	Pass
6	The wind turbine shall rotate within an operating envelope of less than 350 mm.	Measure with an engineering rule	340 mm	Pass
7	The wind turbine shall include a tail to position the blades into the wind, which shall have a surface area of at least 20 000 mm^2.	Calculation based on measurements with an engineering rule	28 484 mm^2	Pass

8	The wind turbine should have 6 blades, each with a minimum surface area of 4800 mm².	Calculation based on measurements with an engineering rule	5180 mm, 5270 mm², 5190 mm, 5255 mm², 5235 mm, 5255 mm²	Pass
9	The blades and tail shall be made from a material which can be recycled.	Check the material certificate	Polypropylene, which is a recyclable thermoplastic	Pass
10	The blades and tail shall be yellow so that they are easily visible from a distance of 5 m.	Visually check from 5 m away	Pass	Pass
11	The body of the shell shall be made of a material which is resistant to corrosion in rain water.	Check the material certificate and materials handbook	Aluminium sheet is listed as being resistant to oxidation by water	Pass
12	There shall be no sharp edges on either the blades or the body of the wind turbine.	Use a silk test to find any sharp edges	No snags	Pass
13	The cost of the parts for the wind turbine shall be less than £20.00.	Adding up the cost of the parts in the material list	The wind turbine was a total of £19.03	Pass
14	The total cost of the wind turbine including labour shall be less than £100.00.	Adding up the material cost from 13 and labour in the production plan (3 hours at £18/hour)	£73.03	Pass

■ Using the evaluation to improve the product

During the evaluation the designers may identify some needs in the specification which have not been satisfied. To address these they may need to propose design changes or changes to the manufacturing processes used to make the product. It may be necessary to go through the design process again and to make another prototype, to test that the product with these improvements will meet all of the needs.

The designers may also use the evaluation to identify how well certain needs were met. Using their understanding of the design principles for the product, they will pay close attention to features or needs where an improvement could result in the product working better or being more attractive to the client or the customer. This can give rise to a continuous cycle of improvement in the product. If the product was a prototype for a larger quantity, then the production will probably still proceed, with a 'new, improved' version of the product being released after it has been made and evaluated.

B *Wind power generator*

Activity

Create a list of the tests that might be carried out during the testing of an electric kettle. Using these tests, work out what the specification for the product was.

Summary

Functional testing is essential to make sure that the product works as intended.

The product should be compared with the specification to check that it meets all of the identified needs, including those that do not relate directly to the function.

If the product does not satisfy all of the needs in the specification, it may need improvements to the design or production process.

Key terms

Functional testing: testing to check that the product does what it is meant to do, sometimes carried out by actual use.

Specification: a list of needs that the product must satisfy.

AQA Examiner's tip

It is important to compare the finished product with the product specification. This can be undertaken with the client or independently.

4

Preparing for production involves selecting the processes and equipment to use and working out how these will be used.

The engineer will look at the design of a product and the materials and equipment that are available to make it. He or she will identify the processes that need to be carried out. These may include materials removal and cutting, shaping, forming, surface finishing and joining and assembling. For each process the engineer will identify suitable tools and equipment to use. There is a large range of different equipment that could be used, depending upon the material being used, the quality and accuracy needed and the cost.

He or she will then decide the order in which to carry out the processes and prepare a production plan. The aims of the production plan are to ensure that the part is produced safely and will meet the needs of the design. It will include the tasks to be carried out, the materials and tools to use, any quality checks needed and safety notes. The production plan will be supported by risk assessments for each piece of equipment that will be used. The engineer may also arrange for the manufacture of jigs, fixtures and templates, if these are required to ensure that the product is made correctly.

Once the preparations are complete, the part will be marked out, if required, and manufactured. Once complete, it will be evaluated to make sure that the needs of the client and customer, as shown in the specification, have been met.

∞ links

Further information on basic workshop tools and equipment
www.technologystudent.com/equip1/equipex1.htm

Further information on machining processes
www.bbc.co.uk/schools/gcsebitesize/design/resistantmaterials/processtechniquesrev1.shtml

Further information on production planning and scheduling, QA and QC and safe working
www.bbc.co.uk/schools/gcsebitesize/design/resistantmaterials/processystemsrev3.shtml

Further information on jigs and fixtures
www.geocities.com/budb3/arts/tools/tjaf.html

Further information on risk assessment and risk management
www.hse.gov.uk/risk/

Further information on safe working practices
www.technologystudent.com/health1/ed1.htm

Further information of the range of health and safety legislation applying to the work place
www.healthyworkinglives.com/advice/minimising-workplace-risks/health-safety-legislation.aspx

AQA Examination-style questions

1. (a) What does CNC stand for? *(1 mark)*

    (b) List two advantages and two disadvantages of using a 2-axis CNC machine for the manufacture of a one-off part, compared with using manual machines. *(4 marks)*

2. List one advantage and one disadvantage of pressure die casting compared with sand casting. *(2 marks)*

3. Describe, with the use of sketches, the steps involved in the vacuum forming process. *(8 marks)*

4. A manufacturer is making a canoe from fibreglass. It will have two parts, a top and a bottom.

    (a) Describe the process that he or she will use to make the shapes of the two halves. *(4 marks)*

    (b) List a suitable method for joining the two parts. *(1 mark)*

5. Name and describe two processes that could be used to join two pieces of mild steel sheet together. *(4 marks)*

6. What is a risk assessment? *(1 mark)*

7. (a) What is the difference between a jig and a fixture? *(1 mark)*

    (b) Describe one of the possible uses of a template during manufacturing. *(1 mark)*

8. What is the name given to the known reference point that is used during marking out? *(1 mark)*

9. What tool should you use to mark out a line parallel to an edge on a mild steel plate? *(1 mark)*

10. (a) What is meant by quality control? *(1 mark)*

    (b) List three considerations when selecting a measurement tool for a quality assurance activity. *(3 marks)*

11. (a) List two items of Personal Protective Equipment. *(2 marks)*

    (b) For each item of PPE identified, give an example of a risk that the use of that item reduces. *(2 marks)*

Double award

12. A manufacturer has made a prototype of the body for a mobile phone using the vacuum forming process. In full production this part will be manufactured using injection moulding. Explain how the process used will affect the design of the manufactured body. *(2 marks)*

13. Mild steel is commonly used to make the body panels for cars. Describe two processes that you could use to improve the corrosion resistance of a sheet of mild steel for this type of application. *(4 marks)*

14. You have been given a sheet of acrylic, of size 1000 × 500 × 5 mm. You need to cut four pieces from this sheet, each of size 250 × 250 × 5 mm. At the exact centre of each square, you then need to make a hole of diameter 10 mm. You must:

    (a) List the tools that you will use to mark out, manufacture and measure the sheets. *(6 marks)*

    (b) List the various operations in a workable sequence. *(4 marks)*

    (c) Prepare a production plan for this operation. Note: this should include the information from (a) and (b) above along with the other information that would normally be included in a production plan. *(10 marks)*

15. Name a process that you could use to provide a durable, corrosion resistant finish on aluminium and describe how this is carried out. *(2 marks)*

16. Describe two methods by which a printed circuit board could be manufactured. *(4 marks)*

17. Explain the difference between quality control and quality assurance. *(2 marks)*

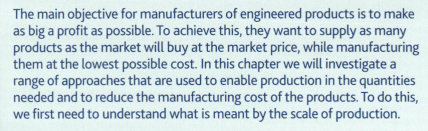

5 Transferring an engineered product to large-scale production

Objectives

List four categories of scale of production.

Explain how the approaches used for different scales of production affect the cost of the product.

The main objective for manufacturers of engineered products is to make as big a profit as possible. To achieve this, they want to supply as many products as the market will buy at the market price, while manufacturing them at the lowest possible cost. In this chapter we will investigate a range of approaches that are used to enable production in the quantities needed and to reduce the manufacturing cost of the products. To do this, we first need to understand what is meant by the scale of production.

Scale of production

There are four commonly used categories for the scale of production. These are compared in Table **A**.

One-off production

One-off production is the manufacture of a single item, usually designed to a customer's individual specification. Examples of one-off products range from a replacement for a broken part in a machine to the engine of an aircraft carrier.

One-off parts are normally made by skilled engineers, either by hand or by using standard workshop equipment. The product costs are very high because of the amount of labour time needed.

Batch production

If a manufacturing company was using batch production to make 300 chairs, it wouldn't make 300 individual chairs one after the other as it would in one-off production. It would make 300 sets of legs, 300 seats etc., then assemble all the parts. This speeds production up considerably.

Batch production is used to make quantities of products ranging from tens to a few thousand. Often these products have similar designs, but are customised in some way. Examples could be parts for Formula One cars or shirts with a designer logo sewn on.

The equipment used for batch production must be flexible so that it can quickly change to the production of a different product. The need to set-up machines means that workers must

A *Characteristics of different scales of production*

	Continuous production	Mass production	Batch production	One-off production
Number of products to make	highest ⟵			⟶ lowest
Cost of equipment	highest ⟵			⟶ lowest
Cost of labour per product	lowest ⟶			⟶ highest
Product cost	lowest ⟶			⟶ highest
Type of equipment used	Dedicated automated machines with fixtures	Dedicated automated machines with jigs and fixtures	Flexible machines with jigs	Flexible machines, hand tools

be skilled. The labour cost is lower than for one-off manufacture as making similar parts together needs less machine set-ups than making products one at a time.

Mass production

Mass produced products include cars, mobile phones and food products. Most things that you experience every day will probably be mass produced.

The cost of setting up a production line for mass production is normally very high. This means that you have to make large quantities of products so that the cost of the equipment can be divided between many products. Most of the machines used will be designed to just do one job, again and again, with only slight changes.

Often the human work force involved in mass production is relatively unskilled because much of the work is automated. The worker may only be adding or removing the components from the machine.

B   *Example of mass production – making cars*

Continuous production

Continuous production is used to make products like steel, oil or chemicals. Many of these products are called 'commodities' and are used to make other products. Factories that operate continuous production often run 24 hours a day, seven days a week. The process needs to be continuous because it would be very expensive to keep shutting down for the night or for breaks etc.

How production systems work in real life

Many products are produced using more than one of these production approaches. A good example is flat pack furniture, such as kitchen units.

C   *Example of a continuous production: making steel strip*

- These units use screws, bolts and hinges which are mass produced.
- The doors and furniture bodies are typically manufactured in batches, as one factory may produce several different styles and finishes.
- When a fitter is putting the unit into your house, he or she may have to make 'one-off' modifications to shape worktops for unusual shaped rooms or cut holes to allow for pipes or electrical fittings.

Starter activities

1   Find five examples of engineered products made using each of the four scales of production.
2   Identify two products that include at least three parts each made at different scales of production.

D   *Example of a kitchen unit*

5.1 Automating processes for quantity production (1)

■ What is CNC?

The instructions to operate most manual machines are fairly straightforward. For example, for a centre lathe:

1 Set chuck speed in revolutions per minute.

2 Turn chuck on.

3 Plan the movement needed in the X, Y or Z axis from the datum (start) point.

4 Set the tool movement speed in mm per minute.

5 Turn the cutting lubricant on, if required.

6 Carry out the tool or job movement.

7 Turn the cutting lubricant off, if used.

8 Return the tool to the datum.

9 Turn chuck off.

Similar instructions are enough to operate a lathe, milling machine, router, drill or other devices. These instructions can be turned into a program that will allow a computer to control a machine, rather than a skilled worker. The computer will read this program as strings of numbers. An example of an actual program is shown in Photo **A**.

Computer Numerical Control (CNC) means using a computer to control a machine using numerical data. All machines that are controlled by a computer are CNC because all machine code is numerical.

■ What are the advantages and disadvantages?

CNC machines have several advantages over manual machines:

- The machining speeds are typically higher, so the parts are machined in less time.
- The surface finish can be more accurate.
- They offer greater consistency, which can otherwise be a challenge when manufacturing products in large numbers.
- CNC machines do not need to take breaks – they can work continually, 24 hours a day, 7 days a week, if needed.
- As noted above, they can carry out several tasks within a single machining operation.
- They can produce shapes that are difficult to manufacture using manual machines, such as freeform curves.
- One operator may be able to mind several CNC machines, further reducing the labour time that goes into each part.

There are two main disadvantages to using CNC machines:

- CNC machines are much more expensive than manual machines.
- It can take a long time to write the program to operate the CNC machine.

Objectives

Explain what is meant by CNC.

Describe the benefits of using CNC machines for large volume manufacture.

List some commonly available types of CNC machines.

A *Program on a CNC centre lathe*

When planning the equipment needed to make a new product, the engineer has to consider the number of parts to be made and the savings in labour time per part, compared with the programming time needed and the extra cost of the machine. This normally means that CNC machines are only used when batches or large quantities of parts are to be made.

■ Types of CNC machine

The use of CNC machines is very widespread. Even within a school workshop, common CNC machines include sticker cutters, routers and mills, lathes and laser cutters. In addition to these, machines used in industry include drills and integrated machining centres, which can carry out both milling and drilling.

Industrial CNC machines can be very complex. For example, a CNC lathe may have more than one chuck. The single tool post on a manual lathe may be replaced by a tool turret that may hold ten or more different tools at the same time. This allows the machine to carry out a lot of different operations one after the other, without loading or separate set-ups. This means that when making a turned part on a lathe, in just one extended operation the part could be taper turned, chamfered, knurled, drilled, bored and parted off, all without human intervention.

Further, these machines can be capable of machining in 2, 3, or 4 axes, Table **B**. As described in Topic 4.2, 2-axis movement can be used to manufacture 'flat' parts. The tool in a 3-axis machine can also move up and down, which allows 3-dimensional parts to be made, Diagram **C**.

B Axes of movements of a CNC machine

No. of axes	Type of movement possible
2	backwards, forwards, right or left on a flat surface
3	backwards, forwards, right, left, up or down
4	backwards, forwards, right, left, up or down. The work piece can also be rotated.

Back   Up   Right
Left
Down   Forward

C 3-axis movement

Information

Development of CNC equipment

1940 First computer

1952 Prototype of the first CNC machine

1954 Programming code developed for CNC

1957 First CNC machines tools become commercially available

Key terms

Computer Numerical Control (CNC): using numerical data to control a machine.

AQA Examiner's tip

You should be able to list three advantages explaining why CNC is used and explain why manufacturers invest in CNC machines.

Activities

1   Using the internet, find pictures of products that were manufactured using 2-axis, 3-axis and 4-axis CNC machines.

2   Write a short article that could be used in a technical magazine, explaining how the use of CNC machines during the manufacture of parts for aircraft could benefit the customer. The article should have a maximum of 200 words.

Summary

Computer numerical control (CNC) means using numerical data to control a machine.

CNC machines can be faster, more consistent and more reliable than manual machines. However, they need time to program and are more expensive.

Commonly available CNC machines include lathes, mills and drills. These may have the capability to use multiple tools and be capable of machining in 2, 3 or 4 axes.

5.2 Automating processes for quantity production (2)

What is materials handling?

When companies invest in CNC machines, one of their main objectives is normally to reduce costs. This is mainly achieved by reducing the amount of labour time, and therefore labour cost, needed to machine and assemble the product. However, this is not the only labour time needed during the manufacture of a product. Time will also be needed for **materials handling**. This means the movement of parts and materials, such as:

- moving parts around the factory, such as unloading and storing materials and preparing finished products for shipment to customers
- moving materials between processes
- loading and unloading parts into machines.

When making large quantities of products, manufacturing companies try to minimise the labour costs per product by selecting the most cost-effective methods of materials handling. As for CNC machines, the engineer will base this decision on a cost analysis. He will consider the number of parts to be made and the savings in labour cost per part, compared with the cost of the equipment. This means that the amount of automated materials handling used is normally directly related to the quantity of parts being made.

Moving materials around the factory

Depending upon the size and weight of the parts, small numbers of parts may be carried by hand, moved on a sack barrow or carried by a forklift truck, Photo **A**. All of these require an operator. As the number of parts, and therefore the total weight increases, the use of sack barrows and forklift trucks reduces the average amount of labour needed per part.

When large quantities of parts need to be moved on a regular basis, Remotely Operated Vehicles (ROVs) may be considered, Photo **B**. These are sometimes called Automated Guided Vehicles (AGVs). They are robot vehicles that can move around a factory without the need for a human driver. Some are controlled by following wires embedded in the floor, others by using lasers which detect their position on the shop floor. ROVs are mainly used in the automotive, aerospace and paper and board industries.

Moving materials between processes

Depending upon the size and weight of the parts, small numbers of parts may be carried by hand, moved on a sack barrow or moved using small overhead cranes. For larger quantities of parts conveyors may be used.

A *A forklift truck*

B *ROVs moving materials*

The simplest form of conveyor is a roller table. This is a series of rollers held between two rails. The products are pushed along by an operator. A small roller table may cost a few hundred pounds.

Belt conveyors move the products on rubber belts, Photo **C**, similar to those used at the check-outs in supermarkets. Slat conveyors are similar, but their belt is made from interlocking plastic slats. They have the advantage that the belt can be opened up for cleaning. In both cases, they are driven by electric motors. The cost of these conveyors depends on their length, but is considerably more than roller conveyors. They are common in the food or pharmaceutical industries.

C   *Conveyor moving food products*

Overhead conveyors are widely used during component manufacture in the automotive industry and for paint spraying applications. They have the advantage of saving space and allowing access underneath the product. They are more expensive than belt conveyors due to the structures required to support them.

The production line used to assemble cars, Photo **D**, is a conveyor. To set up one of these production lines can cost several million pounds. To be cost effective, they need to be used during the manufacture of thousands of cars.

Loading and unloading parts into machines

For small quantities of parts, machines are normally loaded by the operator. For larger quantities, robots can be used to load and unload machines automatically.

D   *Automotive production line*

Activities

1   Using the internet, find examples of materials handling devices used in a range of applications.

2   A manufacturer of steel-framed bicycles has traditionally moved all the products on the shop floor by hand. The company has won new business and is considering a significant increase in production, from 10 bicycles per day to 200 bicycles per day. Write a report explaining how they should adapt their methods of materials handling to be cost effective for the increased volume of production.

Key terms

Materials handling: the movement of parts and materials.

Summary

Materials handling means the movement of parts and materials around the factory, between processes and into and out of machines.

Companies choose the most cost-effective methods for materials handling based on how many parts will be made, the labour cost per part, and the cost of the equipment.

Materials handling methods used for large quantities of products include Remotely Operated Vehicles, conveyors and automated machine loaders.

AQA Examiner's tip

You should be able to explain why manufacturers invest in automated systems for materials handling.

5.3 Using technology to link design and manufacture (1)

What is CAM?

Computer Aided Manufacture (CAM) means using computers to operate machinery to produce a product. CAM is widely used across all sectors of engineering industry.

All machines used for CAM use CNC, as explained in Topics 4.2 and 5.1. This means that CAM has all the benefits of CNC, such as high accuracy, fast machining speeds and, for repeat products, good consistency. Some industrial engineers use the terms CNC and CAM interchangeably.

What is CIM?

The machines used for CAM need to be programmed. If a product has been designed using CAD, computer software can analyse the drawing and create the control program, Photo **A**. This can save hours of manual programming time. This linking of CAD and CAM is called **Computer Integrated Manufacture (CIM)**. CIM combines all the advantages of CAD and CAM. CIM is widely used in companies which design and machine products, including those in the automotive, aerospace, electronics and mechanical engineering sectors.

One important feature that is often overlooked is that the CAD drawing of the finished product is not always suitable to be used directly for machining. It may show features that need to be made on two or three different machines. For example, a part may need turning, then milling, then drilling. Each machine may need separate CAD drawings for use in CIM. These 'stage drawings' can be created by editing the master drawing. This is still much quicker than producing the individual CNC programs.

CIM can be used to reduce the amount of time needed to develop a product and bring it to market. For example, CAD drawings can be sent directly to machines on the shop floor as soon as they are finished. They could also be sent around the world, to companies close to where the product is needed. This means that products can be made where they are needed, saving the time to transport them.

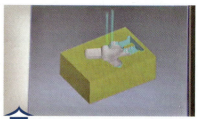

A *A 3-D CAD package creating the machining procedure for a mould*

Case study

Using CIM at a diving equipment manufacturer

One of the products manufactured by a diving equipment specialist is breathing equipment for use by divers. A critical component is the part that the diver puts into their mouth to breathe through, Photo **B**. This part is manufactured using the injection moulding process, where plastic is squeezed into a mould that is the shape of the part.

B *Breathing apparatus for diving*

The part is designed using 3-D CAD software, Photo **C**. A negative image of the part is used to create a CAD model of the mould. CAD software can create this image in seconds. If this had to be drawn by hand it would require a separate drawing, which would take several hours' work. Computer software uses this model to create the control program needed for the CAM machine, Photo **A**.

The mould is then manufactured using CAM, Photo **D**. After checking, the mould is used in the injection moulding machine to manufacture the components, Photo **E**. The finished components are assembled as part of the final product, Photo **F**.

C *Producing a CAD model of the component*

D *Making the mould using a CNC mill*

E *The components leaving the injection moulding machine*

F *The component being assembled into the final product*

Activities

1 Using the facilities available in your school, use CAD to design a simple part and then manufacture it using a CNC machine.

2 Create a cartoon strip that shows the sequence of tasks to be carried out when using CIM to design and make a simple part using a CAM router. This should be suitable for use to explain CIM to Year 7 pupils.

Key terms

Computer Aided Manufacture (CAM): using computers to operate machine tools.

Computer Integrated Manufacturing (CIM): using CAD to design a product, then CAM to manufacture it on CNC machines.

Summary

CAM is the use of computers to operate CNC machine tools.

CIM is the use of CAD to design a part followed by the use of CAM to manufacture it on CNC machines.

CIM combines the benefits of CAD and CAM and reduces the need to write a program for the CNC machine. It can also reduce the amount of time needed to develop a product and bring it to market.

AQA Examiner's tip

When describing CAM and CIM use real life examples, preferably based on a visit or a video you have seen.

Using technology to link design and manufacture (2)

What is CIE?

CIM can be used to manufacture finished parts from CAD drawings. However, there are several other functions that a company may need to carry out in order to supply products. **Computer Integrated Engineering (CIE)** provides a complete business system for a manufacturing company, integrated by a common database. This means that a single CIE system can link together:

- CAD design
- simulated testing of CAD models during the design of products
- the use of software to work out what materials need to be ordered and to place orders on suppliers electronically
- production planning – deciding which products should be made on each machine and the order in which they should be made
- CAM to manufacture products
- keeping quality records
- the tracking and despatch of products using barcodes.

CIE has all the advantages of CIM. It can also reduce the management costs to co-ordinate the different areas of the company and allow savings in material costs. The software can plan what products will be made each day or week. It can therefore calculate what materials are needed to make all of the products within this period. Where the same material is used in more than one product, the full quantity needed can be purchased with just one order. This reduces the number of orders that need to be made for materials, reducing the amount of labour time needed in buying. It can also allow discounts to be obtained for buying in large quantities.

The cost of implementing a full CIE approach can be very expensive. It is mainly used in manufacturing companies where the products contain large numbers of parts, such as the automotive, aerospace, electrical and mechanical sectors.

Concurrent engineering

Concurrent engineering means that different stages of the design process can overlap and be carried out at the same time. For example, this could mean that some of the materials are ordered while some features of the product are still being designed or, for products that contain many parts, some of the parts might be manufactured while others are still being designed and tested.

Concurrent engineering relies on excellent communication between all stages of the design process, for example, design, purchasing, planning, manufacture and distribution. CIE techniques are an important tool that can be used to support this approach.

The benefit of concurrent engineering is that it reduces the amount of time needed to develop a product and bring it to market. This means that new products can be launched before the competitors, which is often more likely to make them successful.

Objectives

Explain what is meant by CIE.

Describe how CIE is used in industry.

Key terms

Computer Integrated Engineering (CIE): using a common database to integrate all the ICT-based activities in a manufacturing company, including CAD, CAM, materials ordering, production planning, etc.

Concurrent engineering: carrying out different stages of the design process at the same time.

Making circuit boards at Syntech Technologies

Syntech Technologies design and manufacture electronic circuit boards for a wide range of customers. Each board may contain up to 366 different electrical components and many thousands in total on each board.

The circuit boards are designed using CAD software, Photo **A**. Based on the components used in the design, and the number of assemblies to be produced, computer software calculates the number of bare printed circuit boards (PCBs) and components needed. This allows the components for many different products to be ordered at the same time. It also means that when the assembly is manufactured the production team can be confident that they will have all of the parts needed.

Producing the boards is a three step process on a production line. First, a machine applies solder paste through a stencil. The stencil is produced using the CAD drawing of the circuit to determine where each component will contact the PCB so that paste is applied in that area only. Next the boards move into a surface mount pick and place machine, Photo **B**. Based on the CAD drawing of the circuit a program determines where each component needs to be placed, Photo **C**. Each pick and place machine positions up to 50 000 components per hour. The circuit boards leave the pick and place machine with the parts placed in the solder paste. Photo **D**. They are then baked in a conveyor reflow oven for around 5 minutes, which melts the solder paste, providing a good electrical connection. Finally, the boards are assembled into the finished products, Photo **E**.

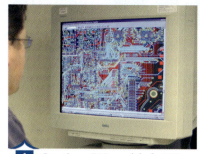

A *Creating a CAD model of a PCB layout*

B *The pick and place machine*

C *Close-up view of the components being mounted on the PCB*

D *The assembled board leaving the pick and place machine*

E *The finished PCB assembled in a product*

Activities

1 Write an advertising brochure that could be used to encourage small machining companies to implement a CIE approach.

2 MRP, MRP2 and ERP are different types of software that are being used by some manufacturing companies. Using the internet, identify what these abbreviations stand for and the main features of each approach. How do these approaches tie in with CIE?

Summary

Computer Integrated Engineering involves integrating all of the business functions of a manufacturing company through a common database.

Concurrent engineering means carrying out different stages of the design process at the same time. It speeds up the process of product development, meaning that products can reach the market in a shorter time.

AQA Examiner's tip

Always emphasise the advantage that computers allow people to work together even though they may be some distance apart.

5.5 Materials selection for large scale production

When selecting materials, the first priority is to meet the functional needs of the product. The product must be able to meet the design needs, irrespective of whether one or 10 000 are being made.

Material properties

For most applications a variety of materials may meet the design needs. When making products in large quantities, instead of choosing the material that offers the best performance, the designer may choose another material that meets the needs, but can be manufactured into the final product more easily.

The **malleability** of a material shows its ability to be formed. This is related to the **strength** of the material and its **ductility**. It can also be affected by temperature and the form of the material. For example, a sheet of polypropylene can be formed over a complex shape when it is heated in a vacuum former, but trying to form the same sheet cold would not be possible.

Forms of material

Materials can be bought in a wide range of shapes and sizes. Common forms range from powder, granules and liquids to sheets, slabs, bars and rods of various sizes, Diagram **A**. An engineer will try to choose a **form** of material that requires the smallest possible amount of work to make the product. This has three benefits:

- It reduces the time needed to make the product. This reduces labour cost and allows more parts to be made within the same time.
- It reduces the cost of materials used, as these are normally related to the weight of the material.
- It reduces waste, as less material has to be removed.

The designer and engineer may choose an alternative to the 'best performing' material if it is available in a more suitable form. For example, consider the metal shaft used in an emergency wind power generator, Photo **B**. The diameter of the broadest section after turning is 8 mm. The material identified by the designer as most suitable was brass, although aluminium could be used. The supplier local to the company making the part has brass rod of 12 mm diameter and aluminium rod of 10 mm diameter. The engineer would probably use the aluminium instead as this would reduce the amount of machining required.

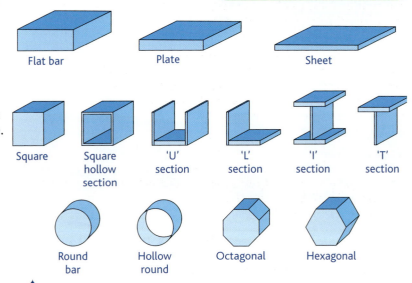

Flat bar   Plate   Sheet

Square   Square hollow section   'U' section   'L' section   'I' section   'T' section

Round bar   Hollow round   Octagonal   Hexagonal

A *Forms that materials are available in*

B *Shaft protruding from the body of the emergency wind power generator*

Consistency and manufacturing

When making large quantities of a product, a key thing that an engineer will look for is consistency of supply. This means that the materials or component used will always be the same. This is important as differences will cause hold-ups in production.

For example, consider a machining operation planned using a 10 mm diameter metal rod. If the supplier ran out and a 12 mm diameter rod had to be used instead, then new jigs and fixtures would have to be made and new programs written for any CNC machines used. Both of these could add significant cost to production.

Similarly, a major advantage of using standard parts such as electrical components is that they are available in the same sizes and with the same characteristics. If, for example, resistors were not available in the same value then electrical circuits manufactured would have to be redesigned every time more resistors were bought.

The engineer will also consider how easily a material or part can be used. This can become very detailed, down to individual manufacturing operations. For example, when using threaded fixings, he or she may prefer cross-head to slot-head screws. This is because it can be easier to locate the screwdriver in a cross-head screw, which means that this operation can be more easily automated. When using metals, the engineer may prefer to use ferrous metals to non-ferrous if he or she has magnetic clamps and lift equipment.

Activities

1   Using the internet, identify the different forms and standard sizes that the following materials are available in: aluminium, stainless steel and polypropylene.

2   Create a table listing the typical strength and ductility of the following materials: low carbon steel, cast iron, stainless steel, aluminium, brass, ABS, nylon, alumina, fibreglass. You could use the internet to find the data.

Key terms

Malleability: the ability of a material to deform under compressive stress.

Strength: how much force a material can resist without breaking.

Ductility: the amount of deformation a material can undergo before it breaks.

Form: the size, shape and condition of a piece of material, product or part.

Summary

When selecting materials for manufacturing products in large quantities, the designer will consider both the properties needed by the product and the ease of manufacturing the product from the possible materials.

The 'best performing' material may not be selected if a suitable alternative is available in a form that is more cost-effective for manufacture.

The material or standard part needs to be consistent, in availability and function, to minimise any disruptions to production.

AQA Examiner's tip

You should be able to explain the practical considerations for the materials used in the product that you make in your coursework project.

5.6 Preparing for production – timing plans

A timing plan has a different role to a production plan. A **production plan** is a complete, detailed set of instructions for making the part. A timing plan is used to plan when a sequence of activities will be carried out. It is a management tool that is also used to monitor whether the activities are on track to finish on time.

One of the most common uses of timing plans is for project management. They are used to manage the construction of new buildings, plan space missions and develop new products. The other main use of timing plans is to plan the manufacture of individual customer orders. In both cases, the format of the timing plan can be similar, Chart **A** and Chart **B**. This format is known as a **Gantt chart**.

■ Why is the timing plan important?

Timing plans allow realistic completion dates to be worked out. In industry, these deadlines will normally be agreed with the customers, or the customers will be advised that they are the delivery dates for their orders. If a company fails to meet its agreed deadlines, the customer may cancel the order. This means that the value of any materials bought or work carried out is lost. Alternatively, some customers will pay less if the products are delivered late. There is also a serious risk that customers may take future business elsewhere.

Time, minutes
1 2 3 4 5 6 7 8

1 Fill kettle
2 Boil the water
3 Put coffee in cup
4 Put sugar in cup
5 Put milk in cup
6 Pour in water
7 Stir

A Gantt chart for making a cup of coffee

■ Creating a Gantt chart

Each step in the project or manufacturing process must be listed in the correct sequence, along with an estimate of how long it will take. Some of the steps will run concurrently – that means they could run at the same time or overlap other tasks. For example, Diagram **A** shows the steps to make a cup of coffee. If you waited until each step finished before starting the next step, it would take a long time to make the coffee. If you put the coffee, sugar and milk in the cup while the water is being heated it is much quicker!

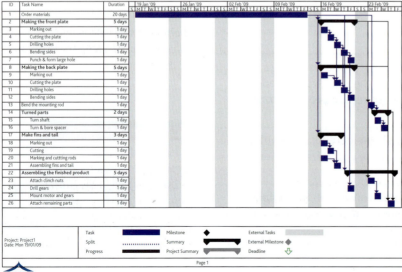

B Gantt chart for the emergency wind power generator

Using timing plans for scheduling

Most engineering companies use batch manufacture. A batch may be part of a single order from a customer, or it could contain parts needed for several orders. A company may have hundreds of customer orders at any one time, all at different stages in the manufacturing process. For every one it will need to work out when the product will be completed, so that it can meet the deadlines agreed with the customers.

Many of these orders may need to be manufactured on the same machines. This creates a complication: to work out the completion date for any product the company would have to take account of every other product being made. Further, every time one product is delayed, re-machined due to quality problems, or machined out of sequence the completion date for every other product could be affected. This would need recalculation, which would take a lot of time. Prior to the development of computers, large companies employed whole teams of people to carry out this activity.

Fortunately, this can now all be calculated using computer production and control systems software. The software can provide a complete list of completion dates. It can also carry out **scheduling**. This means that it will create lists of what needs to be manufactured on every machine in the factory on each day, to achieve the completion dates.

> **Information**
>
> **Using a Gantt chart to work out completion dates**
>
> The shortest possible time in which all the activities can be completed is called the **lead time**. This is the time from the start of the first activity to the end of the last activity, ignoring any overlaps or concurrent activities.
>
> The steps that make up this route through the plan are called the **critical path**. If any of these steps is delayed, then the whole activity will finish later.

Activities

1. Using Diagram **A** for guidance on times, create a Gantt chart for making a cup of tea (with milk and sugar). If you had to make a cup of tea at the same time as the cup of coffee, identify any activities that may clash and re-plan them accordingly. What would be the lead time to make the two drinks separately? What would be the lead time to make the two drinks together?

2. A company has been asked to manufacture a batch of 200 emergency wind power generators. Create a Gantt chart for the part detailed in the production plan in Topic 4.10, Table **C**. What would be the advantages and disadvantages of making the parts in two batches of 100 rather than a single batch?

> **Key terms**
>
> **Production plan:** the instructions on how to manufacture a product.
>
> **Gantt chart:** a type of timing plan.
>
> **Lead time:** the amount of time needed to complete an activity or to supply a product.
>
> **Critical path:** the shortest route through the timing plan, where each step contributes directly to the lead time.
>
> **Scheduling:** deciding when products are being made on each machine.

Summary

A timing plan, such as a Gantt chart, is used to plan when a sequence of activities will be carried out and to monitor progress while the activities are being carried out.

It can be used to calculate realistic completion dates for an activity or the manufacture of a product.

Production planning and control software can be used in a manufacturing company to calculate all the completion dates and carry out scheduling for hundreds of orders. This would take a whole team of people much longer to carry out.

> **AQA Examiner's tip**
>
> Break the processes into small operations and make sure that the sequences make sense.

5.7 Microcontrollers and industrial control systems

Microprocessor systems were first developed in the early 1970s. They are used in a wide range of products and industrial devices, ranging from washing machines and microwave ovens to mobile phones and robots, Photo **A**.

The key component of a microprocessor system is a computer chip that can follow a series of instructions. **Microcontrollers** are a type of microprocessor that can be programmed to control a process or activity. One of the most common types of microcontroller in industry is the Programmable Logic Controller (**PLC**). These can be very robust and suitable for use in demanding environments.

How does a PLC work?

A PLC is effectively a series of programmable switches all included within one microchip. A program is entered into the PLC using a hand-held programmer, a computer or through a wireless network. In operation, the PLC's program continually scans each of its inputs, before using this information to decide whether to activate or change its outputs.

The inputs to a PLC may include switches, such as keypads or buttons. They could also include sensors, for light, temperature, sound, position, liquid level etc.

The outputs from a PLC may go to any electrical device, such as machine motors, lights or buzzers.

The inputs and outputs may be either digital or analogue. **Digital** means that the signal can be on or off, Diagram **B**. For example, consider a simple switch to detect water level in a cooling tank. If the water reaches its level the switch is on; if the water does not reach its level it is off.

Analogue signals can vary continuously in size, Diagram **C**. For example, the signal from a temperature sensor in the cooling tank may increase or decrease on a scale. Analogue signals can be challenging for simple electronic systems to control. However, a PLC can be programmed to carry out an action if the input signal reaches a certain level. For example, in the cooling tank it could be programmed to provide a digital output, turning on a fan if the temperature reaches a certain level. It could also be programmed to provide an analogue output, increasing the speed of the fan for higher temperatures.

A single PLC can handle several different inputs and outputs at the same time. This means it can be used to make complex changes to one process or to control several processes at the same time.

A *Applications of microprocessors*

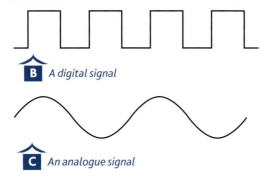

B *A digital signal*

C *An analogue signal*

How are PLCs used?

PLCs are increasingly being used to provide control in a wide range of applications. For example, in the home they are used to control washing machines, dishwashers and microwave ovens. In commercial environments, they can be used to control ticket machines and vending machines.

In industry PLCs are being used in machines to carry out monitoring and control. Here are some examples.

- In the chemical industry, PLCs are used to monitor and adjust the temperature and pressure inside the chemical processing vessels.
- On a production line for chocolate bars, light sensors are used to identify any bar that is misshaped or out of line. The PLC activates a ram to push rejected bars off the line into a waste bin.
- PLCs are also used on a production line to add up the number of parts that have passed a point on a conveyor and group the parts into the quantities to be loaded into each box.

PLCs can also be linked together by a central computer to completely control a production line. This means that every machine on the line is continually monitored, to ensure that the line always works at its optimum level.

Activities

1. Using the internet, find examples of three applications where PLCs are being used to provide process control.
2. If your school has PLC equipment available, create a control program that can be used to monitor and control the process if it goes out of specification.

Summary

Microcontrollers are programmable microprocessors. In effect, they are programmable switches which can turn on or change a number of outputs in response to different inputs.

The most common form of microcontroller is the PLC. These are widely used in industry to provide process control.

Examples of applications where PLCs are used include: washing machines, conveyors and process controllers in chemical plants.

Key terms

Microprocessor: an integrated circuit that can carry out a predetermined program.

Microcontroller: a computer-on-a-chip, containing a processor, memory, and input/output functions.

PLC: Programmable Logic Controller.

Digital: a signal which is either on or off.

Analogue: a signal which can vary in size.

AQA Examiner's tip

You should be able to explain how advancing technology has provided a method of programmable control at affordable cost.

5.8 Robotic systems

What is a robot?

When asked this question, most people think of the androids that appear in science fiction films or machines like that shown in Photo **A**. However, the most widely accepted technical definition of '**robot**' is by the International Organisation for Standardisation, in the standard ISO 8373: 'an automatically controlled, reprogrammable, multipurpose, manipulator programmable in three or more axes, which may be either fixed in place or mobile for use in industrial automation applications'. 'Manipulator' means that it is able to move its object, such as a part or a tool. This means that a wide range of different machines can be characterised as robots:

- Remote-operated vehicles that carry materials through a programmed route around a factory, as explained in Topic 5.2.
- A car that can be driven by its own onboard computer.
- Machines that are programmed to pick up materials and load them into other machines.
- CNC machines which load their own tools, as described in Topic 5.1.
- The automated robot arms used in factories.
- Pick and place machines used to mount parts on circuit boards.
- Mechanical humanoids and androids, such as ASIMO.

Why are robots used?

The first digital, programmable robot, called the Unimate, was used in industry in 1961. It was used to lift hot pieces of metal from a die casting machine and stack them. It had a number of advantages that are shared with most of the robotic systems in use today:

- It could carry out tasks that would be difficult for human workers to do. In this case, one robot could carry out a task that would otherwise need several human workers, due to the weight of the castings.
- It could carry out the task faster than human workers.
- The machine did not need to take breaks; it could work continually, 24 hours a day, 7 days a week.

As microprocessor technology has developed, robots have demonstrated more advantages:

- They can do the same job, again and again, with greater accuracy and consistency than human workers. For example, welding on a car production line, they can make every weld identical in size and position, Photo **B**. Welds carried out by human workers may vary depending upon their skill and how tired they are.
- They can work in conditions that may have safety risks, for example Photo **C**.

Robots have two main disadvantages. Firstly, they are very expensive – to buy a robot arm may cost the equivalent of the total annual salaries of between three and ten skilled workers. Secondly, writing

 A A robot

B *Robotic welding on a car production line*

Information

Why are they called robots?

The word 'robot' was created by Josef Capek in Czechoslovakia. It was first used in 1921 in a play written by his brother, Karl. The play, called 'Rossum's Universal Robots' was about a factory that made artificial workers. This was 40 years before actual robots were invented! It comes from the word 'robota' which means 'hard work' or 'drudgery'.

C *Spray painting on a car production line*

the control programs can take a long time. However, many modern robots can be programmed using a technique called 'play back'. This is where a skilled operator carries out a task whilst complicated sensors tell the robot the series of movements that he or she is making. These movements are copied exactly by the robot. Once the task is complete, the robot can 'play back' the sequence of movements, repeating the task as many times as is needed.

When planning the equipment needed to make a new product, the engineer has to add up the cost of the robots and the programming (or 'play back') time needed and compare these with the number of parts to be made and the savings in labour time per part. This normally means that robots are only used when large quantities of parts are to be made.

Activities

1 Using the internet, find examples of the uses of robots in five different industry sectors.

2 Using card, make a two-dimensional model of a robot arm. It should include some form of gripper and the ability to bend in at least two locations. Work out the 'operating envelope' for your model – that is the area within which it could pick up and move parts.

Key terms

Robot: a reprogrammable manipulator, designed to move parts or tools.

Summary

The term 'robot' refers to a wide range of machines. These are typically programmable manipulators, designed to move parts or tools.

Robots can carry out repeat tasks faster, more accurately and more consistently than humans. They do not need to take breaks and can carry out tasks that would be difficult or hazardous for humans to do. However, they can be very expensive and need to be programmed.

AQA *Examiner's tip*

You should be able to describe at least two operations where it would be better to use robotic systems than people.

5.9 Safely making the product in large quantities

Making the product in large quantities normally involves the use of CNC and automated equipment. There is a common misconception that using this type of equipment reduces the risks to people in a workshop. In fact, the opposite is true, as the potential injuries can be much more severe. However, these risks are reduced through carrying out risk assessments and putting effective preventative actions in place, as explained in Topic 4.11. Unfortunately, every year people die, lose hands or arms or suffer broken bones, because they have failed to follow these actions.

Separating the worker from the hazard

If it is not practical to eliminate a risk completely, an important preventative action is to separate the people in the working area from the hazard.

A serious risk with any machines using rotating parts is that the user may come into contact with the moving part. This may happen either directly or as a result of clothing or long hair becoming entangled with the moving part. To reduce this risk, machine **guards** are used. These provide a barrier between the machine operator and the moving parts. It is a legal requirement that any rotating parts and drive belts must be fenced off to avoid accidental contact. Diagram **A** shows a selection of machine guards.

When using guards, there is a risk that the machine operator may forget to put them in place. It has even been known for machine operators to deliberately leave guards off, so that they can access the machining area. To avoid the serious risk of injury that this causes, most CNC machines use **interlocks**. An interlock is a special type of switch. It prevents the machine's motor from working unless the guards are closed, separating the operator from the hazard.

When a company uses large robotic systems, the robots will move in their planned sequence regardless of whether there are people in the way. Given the size and nature of movement of these machines, it is not normally practical to attach guards to them. For this reason they are normally separated from people completely by being placed in cages, as shown in Photo **B**. The access to the cage is normally interlocked.

Personal protective equipment

It is not always possible to completely separate the machine operators from every risk. For example, they often need to handle a part before and after machining, to load and unload it. In these situations, appropriate Personal Protective Equipment (**PPE**) may be required. Here are some examples.

Objectives

Explain how the use of guards and interlocks can reduce the risks when using automated equipment.

Describe when PPE should be used.

Explain how machine maintenance contributes to safe working.

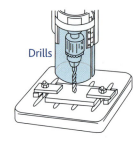
Drills

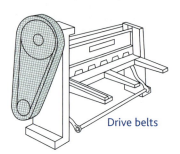
Drive belts

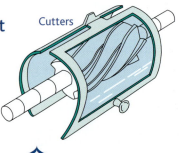
Cutters

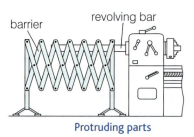
barrier   revolving bar

Protruding parts

A Examples of machine guards

- Goggles reduce the risk of eye injuries from swarf, debris, and broken tools.
- Gloves reduce the risk of hand injuries from sharp edges.
- Aprons reduce the risk of damage to clothes from dirt and debris.
- Face masks reduce the risk of breathing problems from dust particles.

A challenge with PPE is that workshop staff may forget to use it. It should be regarded only as a last line of defence and used only when the associated risk cannot be eliminated using a fixed or permanent solution.

A common mistake is to insist that PPE is used even if there is no associated risk. Its use in inappropriate situations may actually be a hindrance to an operator, increasing the risk of injury.

Maintaining the equipment

Machines are designed to operate safely under normal circumstances. However, if a machine breaks down it may cause a safety hazard. For example, oil or hydraulic fluid could leak on to the floor or into electrical equipment, or a broken component may be a hazard. Machines and equipment need to be regularly serviced by skilled fitters. This not only reduces the risk of a major accident caused by plant failure, it also reduces the chance of a major breakdown leading to a loss in production.

B *Robot on a car production line – note the yellow cage fence around the working area*

Activities

1 For the machines in your school workshop, identify the different types of machine guard. For any CNC machines, identify the interlocks.

2 A trainee machine operator at a manufacturing company has removed a machine guard and short circuited an interlock, so that he can load and unload a CNC lathe faster. The company have discovered this and repaired the equipment. They have decided to give the worker a final written warning. This will explain the reasons why this practice was not acceptable and inform the worker that he will be sacked immediately if he breaks the rules again. Write the letter that the company will give the worker.

Summary

Guards, interlocks and cages should be used to provide a physical barrier between people in the working area and the moving parts of machines.

PPE should be used as a last line of defence when it has not been possible to eliminate a risk through a permanent solution.

Equipment maintenance is necessary to reduce the risk of accidents due to machine breakdowns, as well as stopping disruption to production.

Key terms

Guard: a physical barrier that separates people from an area where there is a risk of injury.

Interlock: a device that prevents a machine operating unless the guards are in place.

PPE: Personal Protective Equipment.

AQA Examiner's tip

You should be able to say why specific precautions or PPE may be needed. A common failing is to suggest that aprons and gloves are suitable PPE for all processes.

Despite the best efforts of manufacturers, most products are not made exactly to the **dimensions** specified by the design, called the **nominal size**. There are many practical reasons for this:

- The tools, moulds and dies used in the machines will start to wear from repeated use.
- There could be variations in the materials or the tools.
- During machining operations, friction from the tools may cause the material to heat up and expand during machining. When it cools again after machining, the part may be slightly too small.
- The machines may vibrate or move slightly during manufacture.
- The tools may slip slightly in the machine jaws.
- The machine or process could have been set up incorrectly.

In practice, most products will have a variation around the size that the manufacturer was aiming for. If every product in a large batch was measured, they would typically have a distribution of sizes represented by a bell curve, Graph **A**. The narrower the bell curve, the more accurate the process is.

Objectives

Explain what is meant by tolerance.

Explain why the dimensions of a part may vary.

Outline some of the actions that could be taken to reduce process variation.

Designers anticipate this variation and for most products they will calculate an acceptable **tolerance**. This is the amount that a component may vary in size from its nominal size and still perform the task that it is needed to do.

The tolerance will depend on the needs of the product and the materials used. For example, the tolerance for engine parts for aeroplanes will be very tight. This is because the performance of the parts is critical and if the metal components need to fit together they have no 'give'. By comparison, the tolerance for the plastic handle of a garden fork will be much larger.

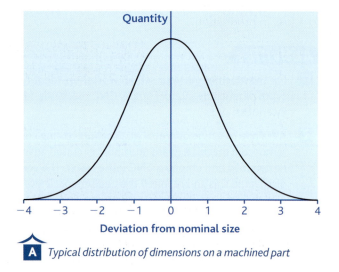

A *Typical distribution of dimensions on a machined part*

Reducing variation in the processes

Manufacturers can reduce the variation using **quality assurance** actions. They can check the composition of the materials as these arrive in the factory. They might change the tools after a set number of uses, whether they look worn or not. They should also carry out regular maintenance on the machines, to ensure optimum performance at all times.

As part of their **quality control**, they can measure the first components manufactured to check that each process is set up correctly. They should also regularly measure further samples of components, to make sure that the machines continue to produce them to the nominal size.

Tolerances for the wheel on a child's scooter

The wheel on a child's scooter is made from injection moulded nylon, with a solid rubber tyre added. It runs on an axle made from a low carbon steel bar, without needing a bearing. The manufacturers do not have the time to try to match pairs of wheels and axles that fit together. They need to ensure that every axle will fit into every wheel.

The nominal size for the diameter of the axle is 10 mm. It cannot be any larger than this, but it could be up to 0.2 mm smaller. This would be written as:

$$10\,\text{mm}\,^{+0.00}_{-0.20}$$

So that the axle will fit, the hole in the wheel must be at least 0.2 mm bigger that the axle. However, if the hole is more than 1.0 mm bigger, the wheel will be too loose, which would affect the performance of the scooter. The 1.0 mm maximum size difference has to take into account the smallest possible diameter of the steel bar, which is 9.8 mm; hence the maximum possible size would be 10.8 mm.

This could be written as: $10\,\text{mm}\,^{+0.20}_{+0.80}$

Often the designer will consider the normal distribution of measurements and change the nominal size of the hole so that it reflects the middle range of the tolerance. This would then mean that the tolerance of the hole would be written as:

$$10.5 \pm 0.3\,\text{mm}$$

Diagram **C** shows how this might appear on an orthographic drawing. If this is achieved, then every axle will fit into every wheel.

B   *Child's scooter*

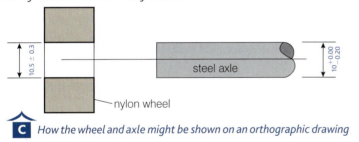

C   *How the wheel and axle might be shown on an orthographic drawing*

Key terms

Dimensions: the sizes of the parts.

Nominal size: the dimension specified on the design.

Tolerance: the amount a dimension can vary from the nominal size without affecting the performance of the product.

Quality assurance: planning and doing things to make sure that the product is made right the first time.

Quality control: checking that the part is correct after it has been made.

Activities

1   Working with three partners, use a rule and scissors to cut a total of 100 pieces of paper. Each piece should be 100 mm long by 10 mm wide. After you have finished measure the lengths using an engineering rule and plot them in the form of a bar chart. What proportion of them were the nominal size? What were the reasons for any variations observed?

2   Using the internet, find an example of an orthographic drawing. Identify the tolerances used for different types of dimension, such as linear distances, angles, radii and diameters.

Summary

Tolerance is the amount that a component may vary in size from its nominal size and still perform the task that it is needed to do.

Variations in the size of manufactured parts can arise as a result of wear, vibration, temperature changes and set up.

Quality assurance actions should be used to try to reduce variations. Quality control can be used to detect variations and correct them.

AQA Examiner's tip

Measure twice, cut once – after you have finished marking out a part, check every measurement again to make sure that they are all correct. This is a lot quicker than having to change a part if it is incorrect.

5.11 Checking the product – measurement

The CNC machines and robotic systems used to make products in quantity are normally more accurate than manual machines. However, it is still necessary to check that the products are within **tolerance**. How much testing is carried out will depend upon the product. For example, an aeroplane engine part will need extensive testing, because a product failure could result in many deaths. If the same approach were used for all products, their cost would rise considerably. A number of alternative approaches have been developed which minimise cost, whilst providing confidence that the product will be within the tolerance.

Objectives

Explain the different approaches to quality control that are used during the manufacture of large quantities of product.

Explain how a Go-NoGo gauge is used.

Computer assisted quality

As explained in Topic 5.7, **PLCs** can be used with appropriate sensors to provide automatic control and adjustment. For example, on an automated production line light sensors could be used to measure the length of every component, with any that fall outside of tolerance being rejected and instructions sent to the process to change its operating parameters accordingly. This approach is very effective, but the equipment is very costly. Therefore it is normally only used where very large quantities of parts are being manufactured or where the part being incorrect might cause safety problems with the product.

Statistical process control

This is based on the principles that the variation of dimensions will normally be in the form of a bell curve and that this curve is well within the tolerance band, Graph **A**. Depending upon the width of the bell curve, it is possible to calculate what proportion of components need to be tested to have statistical confidence that practically all components will be within tolerance. For example, in practice this could mean that out of a hundred components, it may be necessary to measure only six. These measurements are normally carried out by the machine operator or a quality engineer.

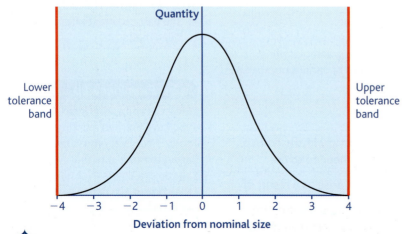

A *Typical distribution of dimensions on a machined part showing tolerance band*

This approach is very effective if the process operates consistently. It significantly reduces the amount of measurement needed and, hence, the cost of measurement. However, if 'special causes' arise, such as variations in the materials or a tool in a machine suffers minor damage, these may not necessarily be detected.

Go-NoGo gauges

A Go-NoGo gauge is an inspection tool used by the machine operator to check every part against the tolerances. It is used to carry out two tests, representing the extremes of tolerance. The part must pass one test (*Go*) and 'fail' the other (*No Go*). It does not produce numerical data, but it does say whether a product is acceptable or unacceptable.

For example, the steel axle of the child's scooter in the previous topic had a diameter of between 9.8 mm and 10.0 mm. Diagram **B** shows a simple gauge that could be used to check this. It is two holes in a piece of tool steel. Every axle would be tested in both holes. If an axle fits into the NoGo hole it would be too small and unacceptable. If it fits into the Go hole it is acceptable. If it does not fit into the Go hole it is too wide and unacceptable.

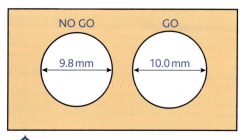

B   *Simple Go-NoGo gauge for bar diameters*

Diagram **C** shows a Go-NoGo gauge that could be used to check that the diameter of a hole in a wheel is between 10.2 mm and 10.8 mm. It is two bars of tool steel. If a wheel fits onto the NoGo bar then the hole would be too large and unacceptable. If it fits onto the Go bar it is acceptable. If it does not fit onto the Go bar the hole diameter is too small and unacceptable.

The advantages of Go-NoGo gauges are that they are quick and simple to use and relatively low cost. The disadvantage is that, although they show whether a product is acceptable or not, they do not generate numerical data which would allow trends to be identified.

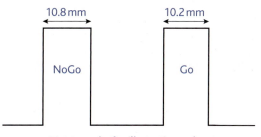

Not to scale: for illustration only

C   *Simple Go-NoGo gauge for the diameter of the hole in the wheel*

Activities

1. Design a Go-NoGo Gauge that could be used to check that the width of a piece of steel is between 20 mm and 22 mm.

2. Using the internet, find out how the number of components to be measured is calculated in Statistical Process Control.

Key terms

Tolerance: the amount a dimension can vary from the nominal size without affecting the performance of the product.

PLC: Programmable Logic Controller.

Summary

Even when using automated machines it is necessary to check that the parts are within tolerance.

Approaches to measurement include measurement by hand, computer-assisted measurement using PLCs, statistical sampling and the use of Go-NoGo gauges.

A Go-NoGo gauge is used to carry out two tests, representing the extremes of tolerance. To be acceptable, the part must pass one test (Go) and 'fail' the other (No Go).

AQA Examiner's tip

For the product you manufacture in your project, you should explain how it could be measured in large scale production. If appropriate, you should include sketches of the designs for any Go-NoGo gauges.

5

When making parts in large volumes, one of the main priorities is normally to minimise cost.

The designer will determine how much the product design can vary from the nominal size on the drawing and still perform the task needed. This tolerance should reduce scrap, by accommodating the practical variations that may arise during manufacturing. The company will source a suitable material, in an appropriate form to minimise the amount of processing required and the amount of waste produced.

The engineer will prepare a timing plan for manufacturing, so that realistic completion dates can be forecast. This will also be used to monitor progress during production. This may be carried out as part of a CIE approach, where all of the operations within the company are linked by a common database.

If it is economic, machining will be carried out using expensive CNC machine tools and robotic systems. These might be programmed directly from the CAD drawings, as part of a CIM approach. The machines will have guards and interlocks, to reduce any safety risks to the machine operators. This type of automated machinery can carry out repeat tasks faster, more accurately and more consistently than humans, without needing to take breaks. The equipment should be regularly maintained, to reduce the risk of accidents and stopping disruption to production due to machine breakdowns. Where necessary, the risks to machine operators will be further reduced by the use of appropriate PPE.

The materials to be machined may be moved around the factory, loaded and unloaded into machines, by other remote-operated vehicles and other robotic systems.

Where economic, computer-assisted measurement will be used for quality control. This may involve the use of microcontrollers and PLCs, to provide instant feedback and correction of any errors. A sample of the parts made will be measured, to minimise the labour cost while providing confidence that all parts should meet the requirements.

∞ links

Further information on scales of manufacture www.technologystudent.com, www.bbc.co.uk/schools/gcsebitesize/design/resistantmaterials/processindpracrev1.shtml, http://tutor2u.net/business/gcse/production_job.htm, http://tutor2u.net/business/gcse/production_batch.htm, www.wisegeek.com/what-is-mass-production.htm and www.spartacus.schoolnet.co.uk/USAmass.htm

Further information on CNC machines
www.technologystudent.com/cam/camex.htm

Further information on the use of computers to assist design and manufacturing
www.gsd.harvard.edu/inside/cadcam/whatis.html and www.cadinschools.org

Further information on microcontrollers
www.technologystudent.com/pics/picdex1.htm, www.howstuffworks.com/microcontroller.htm and www.rev-ed.co.uk/picaxe/

Further information on machine guards and interlocking
www.tdtmachineguarding.co.uk, www.ppma.co.uk/pubs/pdf/MU-2002-November-health-safety.pdf

Further information on safety legislation applying to machines
www.hse.gov.uk/equipment/legislation.htm and of safety legislation generally
www.healthyworkinglives.com/advice/minimising-workplace-risks/health-safety-legislation.aspx

AQA Examination-style questions

Double award

1 A small manufacturing company produces components for use in water pumps. The components are made from metal, using turning, milling and drilling. They have just received a very large order and are considering buying some new CNC machines to manufacture it.

    (a) What is meant by CNC? *(2 marks)*

    (b) Explain three potential advantages of CNC machines over manual machines. *(6 marks)*

    (c) Explain one disadvantage of CNC machines compared with manual machines. *(2 marks)*

2 What is the difference in movement between a 2-axis and 3-axis CNC machine? *(1 mark)*

3 (a) Explain what is meant by CIM. *(3 marks)*

    (b) What is the difference between CIM and CIE? *(3 marks)*

4 A company manufactures custom-designed engines for racing cars. This involves machining operations (turning, milling, drilling, grinding) and assembly. Every customer has a different design. Customers order the engines in batches of up to 20 at a time. At any one time, the company may be working on up to 50 customer orders.

    (a) Give two reasons why they would prepare a timing plan before they start to manufacture an order for a new product from a customer. *(2 marks)*

    (b) How would the timing plan be used during the manufacture of the product? *(1 mark)*

    (c) What are the advantages of using computerised production and control systems software in a batch manufacturing company? *(2 marks)*

5 (a) Name two products that include microcontrollers. *(2 marks)*

    (b) Explain what the microcontroller is used to do in each of the products in part (a). *(2 marks)*

6 PLCs are commonly used for process control in industry.

    (a) What is meant by the term PLC? *(1 mark)*

    (b) Describe an application where a PLC is used in industry. *(2 marks)*

7 Microcontrollers can be used with devices and sensors that provide either digital or analogue inputs.

    (a) State one device or sensor which gives a digital signal. *(1 mark)*

    (b) State one device or sensor which gives an analogue signal. *(1 mark)*

    (c) What is the difference between a digital signal and an analogue signal? *(3 marks)*

8 Robotic systems are used to mount electronic components on circuit boards.

    (a) Compared with human workers, give three advantages of using robotic systems for this application. *(3 marks)*

    (b) State two other examples of the use of robotic systems in manufacturing. *(2 marks)*

9 (a) Give two reasons why it is necessary to use machine guards when manufacturing engineered products. *(2 marks)*

    (b) Explain how an interlock works. *(2 marks)*

10 (a) List three reasons why a product may vary from its nominal size. *(3 marks)*

    (b) Explain two approaches that could be used to ensure that a finished product is still within tolerance. *(4 marks)*

The Controlled Assessment

■ The Controlled Assessment

Units 2 and 4 are practical-based units, where you create a coursework portfolio. Unit 2 applies learning from Unit 1 to practical design and making. Unit 4 (part of the double award) applies learning from Unit 3.

AQA issue a list of the possible activities for Units 2 and 4 each year. Each activity is split into two parts – the design task and the making task.

Your school will tell you what you will do for each part. You might carry out both parts for the same product. This would mean that you design the product first and then make it. Alternately, different products can be used. This would mean that you design a product, but that you will be given drawings for a different type of product for the making task. This approach avoids issues from the design task affecting the mark for making. For example, if your design had any errors or missing dimensions, was too complicated, or needed materials or processes that are not available in your school, these would affect your ability to make the product.

■ Unit 2

The focus of this unit is on designing and making parts as one-offs or in small quantities.

Design task

You will be given a design brief. You need to analysis this in detail, identifying and explaining the key features. As part of this, you will examine several aspects of new technologies and analyse similar products. You will then generate some alternative design ideas and evaluate them. For your preferred solution, you will use a range of standard engineering drawing techniques to communicate the design to other engineers. You will test this idea against the design brief and specification to show how well it meets the needs. You will explain the reasons for your choice and respond to any feedback from the client, modifying your design if necessary.

Making task

Using working drawings of the product and the product specification, you will produce a detailed production plan. This will give detailed instructions for making the product. It will include the processes to use, with explanations for why you would use them. You will make the product, using hand tools, manual machines and computer-controlled machines. You will test the product and compare it with the specification.

■ Unit 4

The focus of this unit is on being able to make parts in large quantities.

This can be a second project, making a different product to Unit 2. However, some activities allow just one finished product to be made, covering the practical elements of both projects. In this case, the project must include at least two different technologies. For example, this means that a combined project would need to include a mechanical and electronic systems, or pneumatic and electronic systems.

Design task

This is similar to Unit 2. The following extra things need to be included:

- During the research and analysis you will need to investigate any energy demands from the product or the manufacturing processes.
- When evaluating the design ideas you will need to carry out objective testing.
- You will need to produce CAD drawings of the proposed solution.

Making task

There are some differences to Unit 2, due to the different focus.

- Your production plan should be suitable for batch and continuous production. It should explain how it differs from a plan for a one-off. It should include methods for quality control and quality assurance.
- The production plan might also include descriptions of how it could be used as part of a Computer Integrated Manufacturing or Computer Integrated Engineering approach.
- You should include several automated or robotic operations.
- You should include reference to, and use of, a 'Smart' material.

> **AQA Examiner's tip**
>
> If you plan your making effectively, this will make it much more likely that your product will achieve the requirements in your specification.
>
> You should make sure that you have practised using the manufacturing processes before using them to make your product.

Glossary

A

Adhesives: compounds used to chemically or physically bond items.

Alloy: a mixture of two or more metals.

Analogue: a signal which can vary in size.

Anodising: the electrolysis of aluminium in an acidic solution.

Assembling: putting parts together.

Assembly drawing: a working drawing showing an assembled product, with all the parts shown.

B

Bending: forming an angle or curve in a single piece of material.

C

Calibrate: checking, and if necessary modifying, a measuring tool to make sure it is measuring accurately.

Casting: the process of making parts by pouring liquid metal into a mould and allowing it to solidify.

Ceramic: an inorganic material, normally an oxide, nitride or carbide of a metal.

Circuit diagram: a schematic diagram using symbols to show how electrical or electronic parts relate to each other in a product.

Client: the company or person for whom the work or design is being carried out.

Communicate ideas: share a concept with others.

Communication: an exchange of views and information between all of the involved parties.

Component: a part used to make up an assembly.

Composite: a material that is made from two or more material types that are not chemically joined.

Computer Aided Design (CAD): the use of computer software to support the design of a product.

Computer Aided Manufacture (CAM): using computers to operate machine tools.

Computer Integrated Engineering (CIE): using a common database to integrate all the ICT-based activities in a manufacturing company, including CAD, CAM, materials ordering, production planning, etc.

Computer Integrated Manufacturing (CIM): using CAD to design a product, then CAM to manufacture it on CNC machines.

Computer Numerical Control (CNC): using numerical data to control a machine.

Concurrent engineering: carrying out different stages of the design process at the same time.

Constraints: things that limit what you can make.

Contractors: another company paid to carry out an activity or make a part.

Conventions: rules of presentation that drawings must conform to.

Crating: using a box to provide guidelines for drawing.

Critical path: the shortest route through the timing plan, where each step contributes directly to the lead time.

D

Design brief: a short statement of what is required.

Design parameters: the values for characteristics that the design has to satisfy.

Die: a mould made of metal.

Digital: a signal which is either on or off.

Dimensions: the sizes of the parts.

Drawing: a visual representation of an idea to communicate detailed information.

Drilling: making holes.

Ductility: the amount of deformation a material can undergo before it breaks.

E

Electrical: circuits containing simple conductors that allow current to flow through them.

Electrical components: devices that are simple conductors that allow current to flow through them.

Electronic: circuits including semi-conductor materials, such as computer chips.

Electronic components: devices including semi-conductor materials in an electrical circuit.

Etching: removal of some of the surface of a part by mechanical or chemical means.

Evaluate: compare how well a product meets the design needs.

Exploded view: a drawing of a disassembled, product showing the components in the correct relationship to each other.

F

Features: details on the design.

Feedback: a response from the client on a proposal or piece of work, which may include changes needed.

Ferrous metal: a metal that contains iron.

Finishing: modifying the surface of the part in a useful way.

Fixture: a work holding or positioning device that is fixed to the machine bed.

Flowchart: a diagram showing a sequence of operations.

Form: the size, shape and condition of a piece of material, product or part.

Forming: changing the size or shape of a material.

Fossil fuels: dead biological material which has been transformed into fuel by geological processes over significant time periods.

Function: what the product is intended to do.

Functional requirements: things that are needed for the product to meet the customers' needs.

Functional testing: testing to check that the product does what it is meant to do, sometimes carried out by actual use.

G

Galvanising: dipping a steel part in molten zinc.

Gantt chart: a type of timing plan.

Globalisation: the process by which companies start operating across the world, rather than locally.

Grinding: using an abrasive wheel to remove a very thin layer of material.

Guard: a physical barrier that separates people from an area where there is a risk of injury.

H

Hazard: something that causes a risk of harm or injury.

Heat: thermal energy.

Hydraulic system: a system that uses fluid to transmit power.

I

Injection moulding: the process of making plastic parts by forcing liquid plastic into a mould and allowing it to solidify.

Interlock: a device that prevents a machine operating unless the guards are in place.

Isometric drawing: a 3-D drawing technique where horizontal lines are at 30° to the horizon.

J

Jig: a work holding or positioning device that is not fixed to the machine bed.

Joining: attaching parts together.

L

Lead time: the amount of time needed to complete an activity or to supply a product.

M

Malleability: the ability of a material to deform under compressive stress.

Materials: the substances or components that products are made from.

Materials handling: the movement of parts and materials.

Materials removal: taking away the sections of the material not required in the final part.

Mechanical: parts used in moving equipment or machinery.

Mechanical parts: moving parts, or parts for use in moving equipment or machinery.

Media: forms used for communication, including images, text, videos, presentations, models etc.

Microcontroller: a computer-on-a-chip, containing a processor, memory and input/output functions.

Microprocessor: an integrated circuit that can carry out a predetermined program.

Milling: using a rotating tool to remove a thin layer of material.

Mitigation action: a precaution taken to reduce a hazard.

Modelling: making a three-dimensional representation of a product.

Mould: a former used to shape a part.

N

Nominal size: the dimension specified on the design.

Non-ferrous metal: a metal that does not contain iron.

Non-renewable: not replenished through natural sources.

Numerical Control (NC): using numerical data to control a machine by electronic means.

O

Operating parameters: the conditions within which the product must operate.

Orthographic drawing: a working drawing of a part showing three views, to communicate the dimensions of the design.

Outsource: to send work to contractors rather than making it in within the company.

P

Painting: applying a liquid which dries to form a coating on the part.

Pattern: a shape used to make a mould.

Performance: how well a product meets the needs of its users.

Photochromic: changes colour with changes in the level of light.

Photovoltaic: generates electric current when exposed to light.

Piezoelectric material: a material which changes shape fractionally when a voltage is applied to it.

Plating: depositing a layer of metal using electrolysis.

PLC: Programmable Logic Controller.

Pneumatic: powered by air.

Pneumatic system: a system powered by air pressure.

Polishing: a physical process where the surface is rubbed or buffed to make it smoother.

Pollution: contamination of the environment.

Polymer: an organic material made up of a chain of single units called monomers.

PPE: Personal Protective Equipment.

Process: an operation that changes the size, shape or condition of a material.

Production plan: the instructions on how to manufacture a product.

Q

Quality assurance: planning and doing things to make sure that the product is made right the first time.

Quality control: checking that the part is correct after it has been made.

Quantifiable: measurable.

R

Recycling: breaking or melting down the material so that it can be used in a new form.

Rendering: applying colour or texture to a sketch or drawing.

Renewable: replenished through natural sources.

Reuse: using the material or component in another application without changing its form.

Risk assessment: a review of the potential of an activity to cause harm.

Rivets: a type of mechanical fixing.

Robot: a reprogrammable manipulator, designed to move parts or tools.

Routing: a method of milling.

S

Sawing: using a blade with sharp teeth to cut material.

Scale: the ratio of the size of the design in the drawing to the size of the part or finished product.

Scheduling: deciding when products are being made on each machine.

Sequencing: putting a series of events in order.

Shading: create different tones on a sketch or drawing.

Shape memory alloy (SMA): a metal which, once deformed, will return to its original shape when heated above its transition temperature.

Shaping: making a part by putting liquid material into a mould.

Shearing: pushing a blade into a sheet of material to separate it.

Sketch: in 3-D CAD, a 2-D drawing produced on a workplane.

Sketching: a quickly produced visual representation of an idea.

Soldering: a method of joining electrical components to circuit boards.

Specification: a list of needs that the product must satisfy.

Strength: how much force a material can resist without breaking.

Sustainable: does not permanently consume resources; can continue indefinitely.

Sustainable materials: materials that are grown or produced as a natural product and can be replaced without permanently consuming resources.

System: a collection of parts that interacts with its environment and performs a function.

Systems diagram: a schematic representation of a system.

Templates: a standard shape that is used as a pattern.

Thermochromic: changes colour with temperature.

Threaded fastenings: mechanical parts such as screws, nuts and bolts.

Tolerance: the amount a dimension can vary from the nominal size without affecting the performance of the product.

Transport: the movement of materials and goods.

Turning: rotating a component at high speed while a cutting tool is pushed into it to remove material.

User needs: the things that the customers require the product to do.

V

Vacuum forming: forming a thermoplastic sheet over a mould, using heat and a vacuum.

Viscoelastic material: material which changes viscosity depending upon the rate at which stress is applied.

Visualise: create an image of what a design will look like.

Welding: a method where a joint is created by melting the contact areas of the parts to be joined.

Working practices: the ways in which people work and companies produce parts.

Workplane: in 3-D CAD, the features are first drawn on a 2-D surface.

Terms used in Controlled Assessment

Component: a discrete assessable element within a controlled assessment/qualification that is not itself formally reported, where the marks are recorded by the awarding body. A component/unit may contain one or more tasks.

Controlled assessment: a form of internal assessment where the control levels are set for each stage of the assessment process: task setting, task taking and task marking.

External assessment: a form of independent assessment in which question papers, assignments and tasks are set by the awarding body, taken under specified conditions (including detailed supervision and duration) and marked by the awarding body.

Mark scheme: a scheme detailing how credit is to be awarded in relation to a particular unit, component or task. A mark scheme normally characterises acceptable answers or levels of response to questions/tasks or parts of questions/tasks and identifies the amount of credit each attracts.

Task: a discrete element of external or controlled assessment that may include examinations, assignments, practical activities and projects.

Task marking: this specifies the way in which credit is awarded for candidates' outcomes. Marking involves the use of mark schemes and/or marking criteria produced by the awarding body.

Task setting: the specification of the assessment requirements. Tasks may be set by awarding bodies and/or teachers, as defined by subject-specific rules. Teacher-set tasks must be developed in line with awarding body specified requirements.

Task taking: the conditions for candidate support and supervision, and the authentication of candidates' work. Task taking may involve different parameters from those used in traditional written examinations, for example, candidates may be allowed supervised access to sources such as the internet.

Unit: the smallest part of a qualification which is formally reported and can be separately certificated. A unit may comprise separately assessed components.

Index

Key terms, and page(s) where defined, appear in bold type.